AN A[...]
YOUR POCKET

Rosemary Ellen Guiley

Thorsons

FOR SUE ROBINSON

ALIS VOLAT PROPRIIS

Thorsons
An Imprint of HarperCollins*Publishers*
77–85 Fulham Palace Road,
Hammersmith, London W6 8JB

The Thorsons website address is: www.thorsons.com

First published as *Angels Among Us* by Pocket Books,
a division of Simon & Schuster Inc.
This revised edition published by Thorsons 1999

3 5 7 9 10 8 6 4 2

A catalogue record for this book
is available from the British Library

ISBN 0 7225 3967 3

Printed and bound in Great Britain by Woolnough
Limited, Irthlingborough, Northamptonshire

ACKNOWLEDGEMENTS

I would like to thank all the persons who granted me interviews for this book, and for their generous sharing of personal experiences. I also would like to thank all the persons who wrote to me to tell me about their experiences as well. I received more stories than I was able to use, and I am sorry that I could not accommodate them all. The angels are busy.

In addition, I'm deeply appreciative to Doreen M. Beauregard for her help in the material concerning the history of beliefs about angels, the images of angels, and guardian angels.

CONTENTS

INTRODUCTION

For many years, I have felt the presence of guiding beings in my life. I cannot pinpoint when my awareness opened to them; rather, it seemed to be a gradual expansion of consciousness that reached out beyond the physical realm. The presence of these beings became more pronounced in adulthood after I embarked in earnest upon my own spiritual quest, both personally and professionally. The more I focussed my attention upon them, the more defined these beings became.

My initial intuitive sense was that these helping beings were 'angels' and that is what I have always called them. The interpretation of nonphysical entities is subjective – what one person calls an angel, another will call, for example, the guiding presence of Jesus, or the Virgin Mary, or God, or Goddess, or the spirit of a beloved one who has died, or an animal spirit guide.

I sense a small group of angels who are around me all the time, connected to my personal and professional lives. They are joined by other angels who come and go depending on circumstances. Perhaps angels, in their individual expressions, have 'specialties' the way humans have in order to make a living.

When I began work on *An Angel in your Pocket*, the angels came out in force. It seemed I had a small army looking over my shoulder to weigh in with their various influences. When I set off to interview someone, my angelic band came with me, and was joined by another band of angels attached to the person I was meeting. Indeed, David Cousins, a clairvoyant and healer whom I met at his home in Cardiff, commented that the room was packed with angels who wished to participate in our conversation.

Angels do much more than guide us in our work. There are angels that look after our relationships, our health, our spiritual growth and our homes. There are angels that come into our lives just to assist us with major transitions. There are angels that accompany us when we travel. Whenever I drive in my car, I visualize 'travelling angels' riding on the hood, bonnet, and roof and clinging to the sides.

Sometimes angels arrive in our lives with a single purpose mission, and depart when that mission is accomplished. When I was going through a dark night of turmoil, which involved the breakup of my marriage, an upheaval in my work, and more than one relocation, I was assisted by angels of healing.

I believe in angels, and I believe that the more we believe in angels, the more they manifest in our lives. That has certainly been my own experience, and the persons who shared their experiences for this book attest to that as well. Angels are ready to help us. All we have to do is acknowledge them.

Rosemary Ellen Guiley

✸

CHAPTER ONE

ENTER ANGELS

✸ Robert owes his life to an angel. While working in a mill, he grabbed hold of a killer 440-volt wire that he didn't know was live. As he felt the electricity surge through his body, he thought, 'My God, I don't want to die!'

✸

Instantly, he was pulled off the wire by an invisible pair of hands around his waist. He was hurled violently backwards and fell to the floor. The next thing he knew, he was lying on his back, looking up at a coworker. He should have been dead.

Robert's colleagues said that his life had been saved by the fact that he was wearing new shoes that day, and the thick rubber soles afforded insulation. Robert believes differently. 'I felt something pull me off,' he said. 'Something saved my life. I thank God I'm alive.' That 'something,' he thinks, was his guardian angel.

Robert suffered some third-degree burns, and lost a little finger to gangrene. Within two years of the accident, he developed psoriasis. Irritated patches of skin appeared around his waist. One day, he noticed that the patches lined up with the imprint of hands. The association was eerie, and reinforced his conviction that divine intervention on the part of one of God's messengers had rescued him from death.

Alice Haggerty too was saved from the brink of death by her guardian angel.

'When I was seven years old, I got sick with diphtheria,' Alice told me. 'I was sick for about ten days. My family was living in a Mennonite community – I was raised Mennonite. We had very strong beliefs about God, the Bible and angels. We did not believe in hospitals or extra medicine – doctors, yes. While I was sick, a doctor came to the house about every day. Medicine was not helping. He wanted to hospitalize me, but it was against our religion.

'I was in and out of delirium from the high fever. But I overheard him one day tell my parents out in the hall that I wasn't going to make it through the night. Strangely, I didn't have any fear about it. I looked forward to going to heaven.

'That night there was a thunderstorm outside. My parents eventually got tired and went to bed. While I was asleep, an angel came into the room. I thought I had died and the angel was going to take me to heaven. He was bigger than a person, and had long hair and flowing robes, and was glowing. I never saw a face. He communicated through my thoughts that he was going to make me better.

'The angel put a silver belt around my waist, picked me up and cradled me in his arms, and took me out the second-floor window, which was closed and had a screen. I could feel a tingling sensation in my body was we went through the window. We went out over a tree. I could see light in the distance. I thought it was the entrance to heaven. We went toward it and entered the light, but I don't know what happened there.

'The next thing I knew, the angel was laying me back down in bed. It was morning, and I woke up. I was soaking wet and the sheets were wet. And I was one hundred percent better.

'The doctor was summoned. He came and examined me and couldn't understand how the diphtheria had disappeared. It was a miracle.'

She had a second experience when her guardian angel appeared to her.

'I was thirteen, and we had just moved to a new house. I was sleeping in bed, in a room with my two sisters. I shared a double bed with my younger sister. I was against a wall. One night something made me jump awake. My sisters stayed asleep. The room had a strange, bright glow. I looked up and there was the angel hovering near the ceiling. I was looking at it, seeing the same long hair and flowing robes. The next thing I knew, I was up there with the angel, looking down at my body in bed. Somehow, this didn't seem unusual to me. I felt the strange tingling again. It seemed like it was quite awhile that I was with the angel, and we were communicating, but I don't know about what. We may have even gone somewhere. Then the next thing I knew, I was jumping back in my body. I looked up, thinking, "Don't go away!" I felt the same awe and sense of beauty and loving feeling that I'd had when the angel healed me, and I felt a tremendous longing to be with the angel.'

For Aimee S. Lacombe her angelic rescuer appeared in the form of a human being. Here is her account:

'I was in a hospital suffering from some rare throat virus that caused me to cough so violently, I would begin to strangle. During one of those fits in the middle of the night, I called for a nurse. No one came right away, and I began to panic, for I couldn't breathe.

'Suddenly the door flew open and a short, stocky nurse came bursting in, and with a voice of authority said, "Close your mouth and breathe through your nose." When I gestured that no air would come through my nose, she clamped her hand over my mouth and shouted, "Breathe!" And, breathe I did and I stopped choking. Her next words were, "Just can't understand why they haven't taught you that." And out she went.

'Because I wanted to thank her, the next morning I asked the nurse who was it who was on night duty.

When she asked me to describe her, she looked puzzled and said that description didn't fit anyone on their staff, but she would check on it.

'Later, the head nurse came in and asked me to describe the nurse again. She said there was no one employed there who came close to my description. When I asked why they hadn't instructed me on what to do when I began to strangle, they said they had never heard of the method.

'The doctor's response to the experience was interesting. He knew about the method, but why he hadn't told me, I'll never know. But he whispered in my ear, "I think you met an angel." By then, I was convinced I had.

'There have been a few times in my life when an almost overpowering aroma of flowers would occur, when no flowers were near. With it was always a sensation of euphoria. I do not think that I am anyone special because of these experiences. But, through them and other spiritual happenings, I know we are here to evolve back to the spirit that we truly are. And once we believe we are not victims of life, but the creators of our reality, all manner of beautiful manifestations appear in our life.'

Robert, Alice and Aimee have plenty of company. In recent years, angels have made a tremendous comeback, and have undergone a resurgence of popularity. More than at any other time in modern history, people are believing in angels, and are talking about their encounters with them. Books on angels proliferate. Not only are people interested in angels, they want to learn how to communicate with God's emissaries.

It wasn't long ago that angels gathered dust, consigned to art and Christmas cards. Except for Catholicism's cult of the guardian angel, most Westerners have scoffed at the idea that angels might be real. Even in the early 1980s, Dutch physician H.C. Moolenburgh found that people laughed at him when he asked if they believed in angels, or had ever encountered an angel. A few people who admitted having encounters with angels were afraid to talk about them, out of fear that others would think they were crazy. Undaunted, Moolenburgh wrote *A Handbook of Angels*. A few years later, the book became an international hit.

Similarly, French journalist Georges Huber encountered a great deal of skepticism about angels when he began research for his book, *My Angel Will Go Before You*, published in 1983. Huber virtually apologizes for his interest in angels, in an age where science and technology make them seem hopelessly out of date and the stuff of fairy tales.

The ecclesiastical world has been almost embarrassed over the question of the existence of angels. Various church figures from various denominations have suggested that angels are 'out of touch with reality' and thus endanger our belief in the Gospels, and that it would be best for all if angels simply disappeared from the liturgy, from sermons – and from public awareness.

Writing in 1969, Cardinal G.M. Garrone, the Archbishop of Toulouse, stated that:

> *'It is an understatement to say that angels have gone out of style. We prefer not to think of them for fear of being confronted with a painful and insoluble dilemma. Either we must affirm with the Church the existence of these mysterious beings and thus find ourselves in the disagreeable company of the naive and uninformed, or else frankly speak out against their existence and be in the equally unpleasant situation of rejecting the faith of the Church and the obvious meaning of the Gospels. The majority, therefore, choose to express no opinion at all.'*

Angels, however, refuse to be consigned to a liturgical waste basket. Popular belief in them will not go away. In fact, belief in angels is at an all-time high, rivalling the level of angel beliefs that peaked during the Middle Ages.

What accounts for this new popularity of angels? There are several major factors. Perhaps the leading factor is a collective sense of lack of control. Every day, we turn on the radio or television and get a litany of bad news. We feel overwhelmed by pressures and circumstances that seem beyond our influence: drugs, crime, homelessness, economic problems, political and social instability, war, disease, famine, and an increasingly toxic environment. We long for help – some sort of divine intervention that, at least if it cannot change things on a large scale, can at least brighten our own little sphere.

Another significant factor in the popularity of angels is that they are an appealing form of divine intervention. Unlike the Judeo-Christian God, who is abstract and has no form or face, angels are personable. According to our mythology, they can assume the form of beautiful humans. They are loving, benevolent, wise, patient and capable of bestowing miracles – or so we perceive them to be (according to Scripture, however, angels will punish humans if that is God's directive.) We view angels as always with us – they never desert us, no matter how poorly we perform. And even though they do not always save us from catastrophe, they stand ready as a source of strength to help us through all our trials. We seem to have a great, collective hunger for spiritual guidance that is personal and intimate, a hunger that is not being met through conventional religion. Angels are our personal companions, our guides, our protectors.

Another major factor in their current popularity is our increasing openness to paranormal experience. Popular interest in the paranormal and things spiritual has gone through cycles in the past. The present interest is part of the so-called Aquarian Age or New Age, which gained momentum around the 1960s.

MICHAEL'S REQUEST

A most unusual angel encounter, which involved the archangel Michael, happened to a dear friend of mine, Juliet Hollister. The encounter is unusual because it featured a vivid visual apparition, clear telepathic communication and some odd synchronicities with other people.

Juliet is founder of the Temple of Understanding, the second oldest interfaith organization in the world. Juliet has travelled the world and has had audiences with, and has been entertained by, an impressive list of heads of state, spiritual leaders and other luminaries.

One might think that Juliet's work with the leaders of the world's major religions would bring her into constant contact with the angel kingdom. Far from it. Many of those leaders with whom she works are openly skeptical of the existence of angels, considering them metaphors or literary devices.

Her earliest encounter with other worlds took place when she was a child growing up in a suburb of New York. Juliet was fond of her grandfather and looked forward every summer to visits at his home. One afternoon in 1920, when Juliet was eight, her mother told her, 'Grandpa has gone to another world.' Two years later, when Juliet once again visited her grandfather's old farm, she looked up from playing in the garden to see Grandpa, looking solid and real as if he were still alive. She received a mental message: 'Surprised, aren't you, little granddaughter. I just popped by to let you know I am keeping an eye on you.' And with that, he was gone.

Juliet told her mother, who dismissed the episode as a trick of afternoon sunlight and a child's imagination. Juliet knew better, and repressed talking about Grandpa's visit.

Juliet's encounter with the archangel Michael took place in October 1984, as she was preparing for the sixth Spirit Summit Conference sponsored by the Temple of

Understanding. The conference was to take place at St. John the Divine, the largest Gothic cathedral in the world, located in Manhattan, New York. The audience was expected to number eight to ten thousand people – a standing room only crowd inside the giant and elegant cathedral.

At the end of the daylong conference, a candlelight ceremony was scheduled to take place. Juliet was to be included in this ceremony, delivering a fifteen-minute talk on the Temple, its purpose and its activities.The prospect of making a speech in front of so many of the world's religious leaders and so many people made Juliet nervous. She slaved away over her speech, and rehearsed it again and again.

The conference was on Sunday, October 7. The final rehearsal was slated for the Friday night before. Juliet travelled to Manhattan on the Friday afternoon and settled into her hotel room. She took a shower and lay down to rest.

As she was getting ready to rise and dress, Juliet suddenly became aware of a presence in the form of a huge column of light standing at the foot of her bed. She perceived the outline of a figure and sensed that it was an angel, though she did not see wings or a face.

Juliet was startled but not frightened. She'd believed in angels all of her life, and knew they didn't show up without good reason – just to check out a hairdo, as she put it later. As she studied it, the presence began to communicate telepathically with her.

'You're going to be speaking at the cathedral, and angels have a lot to do with holy, sacred places,' the angel told her. 'We guard them. Millions of people don't believe in us, but we are real entities. On behalf of the angelic kingdom, we would appreciate it if, when you make your speech, you would tell the people about us, that we are real, that we love the human race, and that we would like to work on behalf of it. But we can't unless we're invited to do so. We don't enter the life of a human unless we're asked. We are very eager to help.'

This message overwhelmed Juliet. Never had she thought of saying anything about angels in her little speech. She replied to the figure of light, 'I really do believe you're an angel. In fact, I kind of think you're the archangel Michael, although there's no reason why I should have one of the top ones show up. I'd love to do anything to help this planet and all the people on it, but I don't altogether trust myself here. I tell you what, I'll make a bargain with you. I'll do it if you do something to confirm that I'm not hallucinating, that I've got this straight.'

The form of light disappeared. Juliet rose, dressed and left the hotel to hail a taxi to go to the cathedral. It was a blowing, cold evening at rush hour on a Friday night. Anyone who has ever been in New York under those conditions knows how difficult it is to find a free taxi. Dozens and dozens roar by, all occupied, no matter where you are in that huge city. And so Juliet stood on the corner waiting in vain for a cab. Fifteen, twenty minutes went by. She grew anxious about arriving on time for the rehearsal.

Then she was struck by what seemed to be a brilliant idea. She said out loud, 'Okay, Michael, here's your chance. I'm in a jam. I've got to get up to the cathedral and I'm running late. Surely in all of Manhattan, you can find one cab that's empty!'

Within minutes, a free taxi pulled up and she hopped in. Now Juliet grew up in New York and had been in hundreds of taxis. She always looked at the dashboard for the driver's photos of his children or wife, or little images of saints, Jesus or the Virgin Mary that are commonly fastened to the dash. But in all her years, she had never seen anything like what greeted her that evening. There stuck onto the dashboard was a cheap plastic statue of a winged form that bore the words, **The Right Archangel Michael.** And it was huge – nearly a foot high.

For a moment, Juliet couldn't speak. This was too weird – even synchronicity seemed unable to explain this incredible 'coincidence.'

The driver was named Tony, according to the license visible on the dash. Finding her voice, Juliet stammered, 'Tony, tell me, what are you doing with the archangel Michael here in your cab?'

Tony turned to her and said, 'Lady, let me tell you, he's a special friend of mine – he's my best friend Mike!'

'Your friend Mike?'

'You don't know about Mike?' Tony asked. 'Hey, he's the greatest! Let me tell you, my wife, she gets mad at me, she throws the spaghetti across the kitchen or whatever, I call on Mike and ask him, how do you handle women. The kids get in trouble in school, I call on Mike. I can't pay the rent, I call on Mike. I really recommend him to you – he can do anything! Of course I have him in my cab. Who else would I have?'

Throughout the journey Juliet heard a lecture on the

virtues of the archangel Michael, and why he was Tony's best and greatest friend. At St. John the Divine, she got out and paid Tony, and he drove off, his statue of Michael standing like a guiding beacon on the dash.

Juliet said to herself, 'All right, Juliet, you asked for a message, and now you've got to keep your promise.' She was determined to keep up her end of the deal, despite some trepidation at the response of ecclesiastical authorities to a message that angels are real.

Inside the cathedral, she asked a docent what angel or angels, guarded it. One graced the roof, but she didn't know its identity.

'The archangel Michael watches over this cathedral,' the docent told her.

The answer was hardly a surprise.

The following night at the conclusion of the conference – a huge success – Juliet said her little piece about how angels are not just pretty Renaissance paintings, but are real, and desire to help humanity. But they cannot do so unless humans ask for their help.

Nobody fainted away in horror at the idea. In fact, for weeks afterward, Juliet received an avalanche of mail, more than she had ever received in her life. The letters were testimony about people's own beliefs and experiences with angels. 'They are real!' was the overall enthusiastic response.

To Test Our Character

Sometimes God sends angels among us, disguised as humans, to test us. The Bible tells us, in Genesis 18, that when Abraham was camped on the plains of Mamre, three men appeared before his tent. He welcomed the strangers, and refreshed them with food and drink. Abraham was informed that Sarah, his wife, would bear a son. The idea seemed preposterous, for both Abraham and Sarah were quite old, and Sarah had never had children. Soon, she bore a son, Isaac.

Ruth Beck was visited by a mysterious stranger one day as she was about to leave her house.

'My three children, ages ten, eight and seven years, were waiting for me in the car. I had to close the doors and turn the lights off. Just as I turned the kitchen light off, I heard a knock on the front door. Thinking it was one of the children in the car playing a prank on me, I opened the door. Standing in front of me, to my surprise, was a very tall, handsome stranger. He was very clean, clothes were pressed, all in black. I took special notice that he wore a cloak with a shoulder-length cape effect.

'He smiled and said, "Could you give me something to eat? I've been on the road two days and have a long way yet to go."

'Startled, I thought, "I'm in a hurry." Then I pondered, "What shall I give him?" There were just two eggs in the fridge. I hurriedly scrambled them, and buttered two slices of bread, and made a nice sandwich. I put greaseproof paper around it and put it in a paper bag.

I added hot coffee to a bit of milk and poured it into a thick white mug, and took it all out to him. He smiled and thanked me.

'Then I turned the lights out and went to join the children in the car. They said, "What took you so long?" I replied, "Why, didn't you see, I fed that man on the porch." They said, "What man?"

'He would have had to pass the car twice to get on the porch. I said, "Let's go look." They hopped out of the car, and we all stood a few paces from the car. It wasn't dark yet. There were no trees, houses or anything to obstruct our view. We could see no one!'

These are but a few of the ways that angels enter our world. No two experiences are ever the same. Yet, we ascribe such experiences to the same agents, angels. What are the factors that have shaped our belief in these beings?

✬

CHAPTER TWO

BETWEEN HEAVEN AND EARTH

✬ Spiritual beings who inhabit a plane of existence between the human and the divine – this is an ancient concept familiar to many religions and cultures.

✬

The angel as it is popularly depicted is a crossbreed, descended from the unearthly entities of Babylonian, Persian, Egyptian, Sumerian, and Greek faiths. Its popular image as a heavenly messenger is generally limited to those monotheistic religions that divide the cosmos into Heaven, Earth, and Hell, requiring couriers to shuttle back and forth between the divisions. This particular brand of angel originated in Persia's Zoroastrian faith and was then handed down to Judaism, Christianity and Islam.

The word angel itself is a mutation of the Greek *angelos*, a translation of the Persian word 'angaros,' or 'courier.' The Hebrew term is *mal'akh*, meaning 'messenger' or 'envoy.' Even this definition is ambiguous, for 'messenger' or 'envoy' is used in five different senses in Scripture. It can mean:

1 *the Word, given by God to the world*
2 *St. John the Baptist, the precursor of the messiah*

3 *priests, who act as God's ambassadors to people*
4 *prophets*
5 *angels*

Generally, the term *messenger* or *envoy* is used in Scripture to mean angels. But messenger is only one of their functions, and these enlightened spirits can be found worldwide, throughout history, playing many roles.

HISTORY OF THE ANGEL

We can only guess at the age of the angel. Images of supernatural winged creatures have been found in ancient Mesopotamia and Sumeria. The Assyrians had their *karibu* (the source of the word 'cherubim'), which were fierce, winged beasts possessing features both animal and human. The role of the angel as protector can perhaps be traced to these ancestors, which acted as temple guards in Babylon and Sumeria.

The Greeks made a big contribution to angel lore with their gods, such as Hermes, the winged messenger. Hermes is often credited as being the source of archangel Michael. (Many of the Greek gods were molded into angels by the Church in its attempt to convert the pagans.) The Greeks also had *daimones*, spirits who came in both good and evil forms, the good ones being protectors. Socrates spoke of his daimon, who constantly whispered in his ear. Daimones evolved into 'demons' in Christianization, and in the process they lost their good-natured brethren.

The Aryans who came to India and Persia around 2500 B.C. believed in *devas* (meaning 'shining ones'), who were deities subordinate to their supreme god, Dyeus. Perhaps it was from them that angels inherited their most salient characteristic – the ability to shine, or radiate light. The 'el' suffix so common in angels' names is understood in several languages to mean 'shining' or 'radiant.'

The devas made their way into the *Veda*, a collection of early sacred Hindu writings, where they were depicted in a hierarchical (but still polytheistic) arrangement. According the Veda, devas existed in the three worlds – Earth, Heaven and a spiritual realm in between. They were closely aligned with the elements of nature – fire, water, earth and air – which were considered expressions of their existence. Devas of water, for example, were assigned the feminine role of caretakers, or nourishers, of all living things.

Devas also found their way into Zoroastrianism, the religion founded by the prophet Zoroaster (Zarathustra) in sixth-century Persia. It was through Zoroastrianism that devas evolved into angels. In founding this monotheistic faith, Zoroaster rejected the pantheism of the Hindus and offered instead a single, supreme deity, Ahura Mazda, locked in an eternal struggle against his evil enemy Ahriman. Ahura Mazda is aided in this struggle by the good deeds of humans. He is also aided by seven archangels, the *amesha spenta*, who are the gods of Babylon and Assyria recycled into more roles more appropriate to a monotheistic religion. They represent the concepts of wisdom, truth, immortality, deserved good luck, piety, salvation, and obedience.

Zoroaster's brand of angels took hold and was handed down to Judaism, Christianity and finally Islam. Islam's *malaika* (again, 'messengers') are androgynous beings made of light who act as guardians of humans. Their names and personalities are borrowed from Judeo-Christian angels – for example, Mika'il (Michael) and Djibril (Gabriel).

JUDEO-CHRISTIAN ANGELS

Zoroastrianism was popular during the Hebrews' captivity in Persia and made its mark on their biblical writings. While their religion did not allow for any gods other than Yahweh, the idea of angelic intermediaries to the

Divinity was acceptable and adoptable.

Angels appear as guides to heaven in the Jewish Kabbalah, secret writings containing mystical knowledge that is passed on to those advanced enough to comprehend it. There are also beings called *sefirot*, which are composed of the energy of God and are known as Grace, Wisdom, Splendour, Understanding, Knowledge, Foundation, Eternity, Power, Beauty, and Crown. They stand together to form a tree – the Tree of Life.

At the top of the tree is the Angel of the Lord, Metatron. He is considered in some accounts to be the greatest of the angels – there is scarcely an angelic duty or function that is not related to him. Metatron also is the representative of God who led the tribes of Israel through the wilderness, and probably is the dark angel who wrestled with Jacob. He is sometimes identified with Satan. In Judaic literature, the principal name of Satan is Samael, which means 'poison angel.'

The etymology of Metatron is unclear. Possibly, the name itself was intended to be a secret, and may have been produced through a glossolalia-type of altered state of consciousness. Glossolalia is speaking in tongues, and is perhaps best-known for its part in charismatic religions. Metatron is sometimes called the Prince of the Countenance, meaning he is the chief angel of those angels who are privileged to look upon the countenance of God.

Descriptions of him tell of a spirit or pillar of fire with thirty-six pairs of wings and myriad eyes. His face is more dazzling than the sun. He serves as God's Angel of Death, instructing Gabriel and Samael which human souls to take at any given moment.

ENOCH'S SEVEN HEAVENS

We might think that the Bible would inform us about the origin, nature and functions of angels, but it falls far short of the task. In fact, the Bible is rather vague on the subject, mentioning angels or alluding to them without offering much in the way of specifics. Only two angels are mentioned by name: Michael ('who is as God'), and Gabriel ('God is my strength'). The Apocrypha mentions others, most notably Raphael ('the shining one who heals') and Uriel ('fire of God'). According to the Book of Revelation, in the New Testament, there are seven archangels who stand before God, but we are left to guess at their identities.

Most of the angel lore that has been handed down through the centuries comes from the Apocrypha ('hidden books'), writings that were declared heretical by the Church for one reason or another. The Apocryphal Book of Tobit tells a great deal about the archangel Raphael. It is the Chronicles of Enoch that tell us the most about angels in general.

According to legend, Enoch was a prophet who was so spiritually advanced that God had angels take him

directly to heaven to record what he saw. Enoch recorded his travels there in detail. His experiences enlightened
him so much that God turned him into the angel
Metatron.

Most likely, the Chronicles of Enoch were written by
numerous anonymous authors around the first and second centuries A.D. The Chronicles were declared apocryphal in the fourth century A.D. by Saint Jerome. One
of the chief objections to them was that they portray a
multi-layered heaven that contains hell, populated by
fallen angels. The concept of hell within heaven apparently was too much for early Church authorities.

Enoch's accounts are compelling reading. According
to the story, Enoch was 365 years old when two angels
appeared and whisked him off to heaven.

Enoch remained in heaven for sixty days. He recorded what he saw in 366 books, which were handed down
to his sons. After a hiatus back on earth of thirty days,
Enoch was once again transported to heaven, where God

immortalized him as Metatron, installing him in the seventh heaven.

Enoch's heavenly adventures began one day when he was asleep on his couch. A 'great distress' came into his heart, which he could not understand. Two beings appeared before him, which he described as men:

> And there appeared to me two men, exceedingly big, so that I never saw such on earth; their faces were shining like the sun, their eyes too were like a burning light, and from their lips was fire coming forth with clothing and singing of various kinds in appearance purple, their wings were brighter than gold, their hands whiter than snow.

Enoch was frightened, but the angels assured him to fear not – they had been sent by God to take him into heaven. They bore him up on their wings to the clouds to the first heaven. In all, Enoch found seven heavens and seven corresponding earths, all united to each other by hooks. Beyond the seventh heaven lay three more heavens.

The first heaven, ruled by Gabriel, is the one closest to Earth, and contains the winds and the clouds. The angels who live here are astronomers and rule the stars and heavenly bodies. There also are angels who are guardians of ice, snow, flowers and dew.

The second heaven is ruled by Raphael and is a dark penal area where fallen angels await judgment. His guiding angels, said Enoch, 'showed me darkness, and there I saw prisoners hanging, watched, awaiting the great and boundless judgment, and these angels were dark-looking, more than earthly darkness, and incessantly making weeping through all hours.'

Enoch was told that these prisoners were 'God's apostates,' fallen angels who had turned away from God with their prince, who was fastened in the fifth heaven.

The third heaven is ruled by Anahel and is a land of

contrasts. One part of this heaven, the northern section, is actually hell, ice cold and sulfurous, filled with torturing angels who punish the evil souls who reside there. The condemned include those who dishonor God, sin against nature, and practice enchantments and witchcraft. However, the rest of this heaven is an Eden-like garden where manna is made and the souls of the holy – those who are righteous and compassionate – reside. Angels of light watch over this heaven.

The fourth heaven, under the jurisdiction of archangel Michael, contains Holy Jerusalem and its temple, all made of gold, surrounded by rivers made of milk, honey, wine and oil. The Tree of Life is to be found in this heaven, as well as the sun and the moon. Here Enoch heard the singing of angels:

In the midst of the heavens I saw armed soldiers, serving the Lord, with tympana and organs, with incessant voice, with sweet voice, with sweet and incessant voice and various singing, which is impossible to describe, and which astonishes every mind, so wonderful and marvelous is the singing of those angels, and I was delighted listening to it.

The fifth heaven is yet another prison, a fiery ravine where the angelic Watchers, or Grigori, are being imprisoned for marrying into the human race.

The sixth heaven is full of scholarly angels, studying astronomy, nature, and *homo sapiens*. Here Enoch found archangels, as well as angels who rule over all the cycles and functions of nature – the seasons, the calendar, the rivers and seas, the fruits of the earth, the grass, etc. There also are angels who record all the lives and deeds of every human soul, to set forth before God.

In the seventh heaven, Enoch found the higher angels, such as the thrones, cherubim, seraphim and dominions, as well as 'fiery troops of great archangels.' These angelic hosts bowed down before the Lord, singing his praises.

Enoch's two angel guides left him at the end of the seventh heaven. He was terrified, but Gabriel appeared before him, caught him up as though a leaf caught up by wind, and transported him higher.

In the eighth heaven, called Muzaloth, Enoch saw the 'changer of the seasons, of drought, and of wet, and of the twelve signs of the zodiac.' The ninth heaven, called Kuchavim, holds the heavenly homes of the twelve signs of the zodiac.

Then archangel Michael escorted Enoch to the tenth heaven, where he held the face of God. It was, said Enoch:

> ...like iron made to glow in fire, and brought out, emitting sparks, and it burns.
>
> Thus I saw the Lord's face, but the Lord's face is ineffable, marvelous and very awful, and very, very terrible...

And I cannot tell the quantity of his many instructions, and various voices, the Lord's throne very great and not made with hands, nor the quantity of those standing round him, troops of cherubim and seraphim, nor their incessant singing, nor his immutable beauty, and who shall tell of the ineffable greatness of his glory?

God instructed Michael to 'take Enoch from out of his earthly garments, and anoint him with my sweet ointment, and put him into the garments of my Glory.' Enoch then took on a shining appearance, like the angels around him. God summoned the archangel Pravuil, 'whose knowledge was quicker in wisdom than the other archangels, who wrote all the deeds of the Lord,' to bring out the books of knowledge for Enoch to read.

After instruction from Pravuil, Enoch wrote his 366 books in thirty days. He was then summoned by the

Lord, who revealed to him the creation story, and that the world would end on the eighth day of creation (or after 8,000 years), when time would cease. Enoch was given all the rules of morality and righteousness for humans to live by.

After these revelations, Enoch was sent back to earth for thirty days, so that he could pass on the teachings to his sons and others. God then took Enoch back into heaven.

THE ANGELIC HIERARCHY

Just as the Judeo-Christian concept of the cosmos is hierarchical, so is the arrangement of angelic beings that inhabit it. But the borders seem to blur, as angels sometimes occupy more than one position and play multiple roles simultaneously. For example, before the Fall, Lucifer was the ruler of the seraphim, the highest order of angels; at the same time, he was also considered an archangel, a position much lower on the totem pole.

The best-known hierarchy derives from the two main sources of information on the subject: Pseudo-Dionysius's *Celestial Hierarchies* and Saint Thomas Aquinas' *Summa Theologica*.

Pseudo-Dionysius, also known as Dionysius the Areopagite, is a mysterious Greek figure credited with authoring some rather dense but enduring theological and philosophical tracts. The name is a pseudonym, and the real identity of the person – or persons – writing under it is unknown. Moreover, the exact dates of his writings are unknown as well, although it is believed that they were done in the fifth or sixth centuries A.D. The works were influential on medieval Christian thought, including Saint Thomas Aquinas (1225–1274), one of the great theologians of the Middle Ages.

Whoever he was, Pseudo-Dionysius followed the Platonic and Neoplatonic pattern of three. He conceived of all reality in triadic hierarchies. In *Celestial Hierarchies*, he presents an angelic kingdom divided into three realms, each of which contains three orders, each of which further is broken down into three levels of intelligences. In each group of three, the function of the first level is perfection or union. Of the second level, it is illumination, and of the third level, it is purification. This triadic arrangement provides a means for the divine, or spirit, to descend into matter, and for matter to ascend to divinity.

A hierarchy, said Pseudo-Dionysius, is 'a sacred order, a state of understanding and an activity approximating as closely as possible to the divine.' The goal of a hierarchy 'is to enable beings to be as like as possible to God and to be at one with him.' Thus, through the angelic hierarchy, humankind has a way of reaching up to God and sharing in the Light that emanates from him.

Pseudo-Dionysius said that angels are nearer to God than are humans, 'since their participation in him takes so many forms.' He elaborated:

Their thinking processes imitate the divine. They look on the divine likeness with a transcendent eye. They model their intellects on him. Hence it is natural for them to enter into a more generous communion with the Deity, because they are forever marching towards the heights, because, as permitted, they are drawn to a concentration of an unfailing love for God, because they immaterially receive undiluted the original enlightenment, and because, ordered by such enlightenment, theirs is a life of total intelligence.

At the center of the triadic angel kingdom is God, emitting rapid vibrations which slow as they move out from the source and become light. Nine orders of celestial beings surround God, divided, as mentioned, into three triads. At any level, the higher angels possess all the powers of those lower, but not those of any higher level.

The upper triad is composed of seraphim (love), cherubim (knowledge), and thrones. The entities of the upper triad are in direct contact with God. They receive 'the primal theophanies and perfections,' said Pseudo-Dionysius, which they pass on to the middle triad. The middle triad consists of dominions, powers and virtues. They in turn pass the word to the last triad, which is home to principalities, archangels, and finally angels. This last triad is responsible for communicating God's word to humans.

THE LEVELS OF ANGELS ARE:

Seraphim. Their name means 'fire-makers' or 'carriers of warmth.' The seraphim set up the vibration of love, which in turn creates the field of life. These angels are of such subtlety that they rarely are perceived by human consciousness. When they so choose, they can manifest in a human-like form with four heads and six wings –

two for flying, two to cover the face and two to cover the feet. They are able to shed their skin and appear in their true, brilliant form. For this reason, seraphim are often identified with the serpent or dragon.

Enoch declared that there are only four seraphim, one for each direction.

Cherubim. The name comes from the Hebrew 'kerub,' which means either 'fullness of knowledge' or 'one who intercedes.' Their name, said Pseudo-Dionysius, 'signifies the power to know and see God, to receive the greatest gifts of his light, to contemplate the divine splendor in primordial power, to be filled with the gifts that bring wisdom and to share these generously with subordinates as a part of the beneficent outpouring of wisdom.'

Cherubs have four faces and four wings. They also function as charioteers to God.

Thrones. Also known as 'ophanim' and 'galgallin,' these

creatures function as the actual chariots driven by the cherubs. They are depicted as great wheels containing many eyes, and reside in the area of the cosmos where material form begins to take shape.

Dominions. Dominions (also known as 'dominations,' 'lords' and 'kuriotetes'), according to Pseudo-Dionysius, regulate the duties of the angels. According to others, they function as 'channels of mercy.'

Powers. Also called 'potentiates,' 'dynamis' and 'authorities,' these angels act as patrols, on the watch for demons trying to exert their evil influence. While they work to maintain a balance between good and evil, they themselves can be either good or evil.

Virtues. These beings are in charge of miracles and providing courage, and are sometimes referred to as 'the shining ones' or the 'brilliant ones.'

Principalities. These beings have 'princely powers' and watch over the actions of the Earth's nations and cities. They are also in charge of religion on this planet.

Archangels. Archangels are a busy lot: they are liaisons between God and mortals; they are in charge of Heaven's armies in the battle against hell; and they are the supervisors of the guardian angels.

Michael is the most famous of the archangels, and the only one to be named an archangel in the Bible. He is said to be the angel who defeats Satan and hurls him into Hell, and who will return with the keys to Hell in order to lock him up for one thousand years.

Other angels have been accorded the rank of archangel. Gabriel is the angel of annunciation, resurrection, mercy and death, and is the angel who appeared to Mary to inform her of her unusual pregnancy. Gabriel is said to guide the soul from paradise to the womb and there instruct it for the nine months prior to birth.

Raphael's name originates from the Hebrew *rapha*, which means healer or doctor; thus Rapha-el is 'the shining one who heals.' As such, he is often connected with the symbol of healing, the serpent. He is entrusted with the physical well-being of the earth and its human inhabitants, and is said to be the friendliest of the angels.

Uriel is a punishing angel – the one who stands with fiery sword at the gates of Eden. He also punished Moses for not circumcising his son. But Uriel is not without a compassionate side. He is said to have given the Kabbalah to humanity, and to have warned Noah of the coming flood.

Angels. The class of higher beings closest to the material world are the angels. They are the true couriers of the heavens, delivering the messages of God to humans. Angels also have a reputation as the musicians of the cosmos. The vibrations of their music are more spiritual impulses than sound waves; thus it is not heard by humans as music but experienced as moods and inspirations.

REBEL ANGELS

To err may be human, but apparently we aren't the only ones capable of making bad decisions. Most angel lore contains stories of angels gone bad. According to the Bible, one-third of Heaven's legion fell from grace – figuratively and according to some legends, literally.

Enoch witnessed a penal colony in the fifth heaven where gigantic angels known as the Grigori (or 'Watchers', or 'Those Who Sleep Not') are kept prisoner. As the story goes, about two hundred of the Watchers were living up to their name – watching over the Earth – and found themselves watching the female descendants of Adam and Eve with more than brotherly love on their minds. They finally decided to disobey God, descend to Earth, and take the women as wives. This act had many problematic consequences. It was a sin of hubris – a mixture of pride and lust. It also resulted in a mixing of the spiritual with the material, which was strictly forbidden as far as God was concerned. And the Watchers imparted to their wives angelic knowledge not meant to be shared with humans – secrets on growing herbs, working with minerals, working 'enchantments,' etc.

Worse still, the union between the fallen angels and the daughters of Eve resulted in very unnatural offspring. These nephilim, or 'fallen ones,' were gigantic, destructive creatures. They required great amounts of food, and when it ran out they would eat humans and even each other. The purpose of the great flood was to eliminate these abnormal beings and preserve the human race. However, there was a second occurrence of these creatures in Canaan, and Israel was charged with destroying them. The destruction was incomplete, and many got away, fate unknown.

The fate of the Watchers themselves was to be held prisoner forever. Their leader, Azazel, was cast by Raphael into a dark pit (either in the Earth or in the fifth heaven) and remains there until Judgment Day, when he will be cast into the fire.

The most infamous of the rebel angels was, of course, Lucifer, who is usually credited with leading the other angels in the big Fall. Lucifer is erroneously identified as Satan (from the heavenly office of *ha-satan*, or adversary), who, while still in the heavens, appeared to hold more than one position in the hierarchy simultaneously. Similarly, he possesses a multitude of aliases and personalities – Metatron is one. But most share the same basic life story: the most glorified of the angels, most beloved of God, made a colossal blunder of pride and was condemned forever to a dark, embittered existence. He lost his luminous quality and inherited the goat-like features of his randy Greek ancestor, Pan.

Perhaps it was Lucifer's refusal to bow to Adam when God presented him to the angels. One legend has it that Lucifer refused to bow because he had been previously instructed to bow to no one but God. He was just obeying orders. Perhaps he refused to bow to any but God because of his deep love for the Creator. Or perhaps he was simply jealous of God's creation, which would explain why he's spent his entire post-Fall life trying to separate humankind from God. Regardless, he was flung into Hell, and one-third of the angels followed him.

GNOSTIC ANGELOLOGY

The Gnostics were a mystical Christian sect that flourished in the second century. They believed that matter is evil and that freedom comes from gnosis, or spiritual truth. The Church considered them heretical, and persecuted and purged them accordingly.

One of their acts of heresy was in disagreeing over the Judeo-Christian theory of creation. Gnostics said there is no singular, higher being who created the world. Rather there is an all encompassing divine spirit which divided itself, creating various worlds and beings, human and spiritual, in the process. One of these was the principal, Sophia, who represented the concept og wisdom. Other emanations from the source included the concepts of love, power, truth, and mercy. These formless concepts eventually solidified into angelic beings.

At one point, the divine source withdrew itself from these beings and gave them the free will to stay or to go. Some chose to remain close to the source, serving it and aiding those humans who desire reunion with the divine. Others – the 'fallen' angels – became so alienated from the source that they grew arrogant, all claiming themselves to be the One and Only. According to the Gnostics, the Creator who is worshipped in Judeo-Christian religions is actually one of these arrogant fallen angels.

Satan falsely claimed to be the creator and ruler of the cosmos, and we believed him.

ANGELIC BEINGS IN OTHER RELIGIONS

Even religions and cultures that do not embrace the Judeo-Christian concept of angels believe in spiritual beings who act as mentors and guardians.

While the angels of Western religions are believed to be a species genetically distinct from humans, the divine beings of Eastern religions are often humans who have reached a high rung on the karmic ladder. Buddhists and Hindus do not believe in angels as bearers of great truths; rather, they have wise men who fulfill this function. Buddhist devas include mortal beings who have earned a place in the astral plane not accessible by regular human beings; however, being mortal, they are subject to the laws of reincarnation.

In Mahayana Buddhism the 'bodhisattva' are one step beyond the enlightened bodhi of Buddhism. Bodhisattva are humans who have attained not only enlightenment but also total compassion. They reside between Heaven and Earth, and like the deva, must also pass through the cycle of reincarnation. The spiritual energy within them is referred to as a *dakini*, or angel. Bodhisattva act in the usual angelic capacity toward humans, teaching and healing; they also guide the spirits of the dying.

The Taoists have their Immortals, which are former humans who have attained physical immortality and status as demi-gods. They fulfill the usual angelic duties of healing, protecting and delivering messages. The Hindu believe in birdlike *apsaras* which carry the dead into paradise. And there are the *kinpuru'sh*; these are winged creatures that remain close to God at all times.

Candomblé (a form of Santería) is a Brazilian religion, adopted from Africa, that emphasizes communion with nature as the path to guidance and enlightenment. According to this faith, we can receive truth and wisdom by approaching the spirits of nature with an attitude of humility and good intentions. The angels of this religion are known as *orishas*. Orishas are immaterial beings who act as helpers to Olorun, the god of Candomblé, and each is aligned with an element of nature, much like the Hindu devas.

At birth everyone receives a male and a female orisha, which behave in the manner of guardian angels. In order to discover the identities of one's orishas and receive their guidance, a person must go to a 'power spot,' a place in nature where one feels most at peace and attuned to the cosmos. When a person intuitively discovers power spots, he or she discovers the home of the personal orishas, and it is there that their guidance can best be received.

Native Americans have angel-like creatures in animal form, spirits who sometimes help people and deliver messages from above. The shamans, in their visits to the underworld, are accompanied by animal spirit guides. And many Indian cultures believe that the spirits of their ancestors act as guardian angels to the living.

The Celtic peoples have their fairies, sprites, elves, brownies and leprechauns. The stories of their origin are many, but one legend claims that they are the angels who fell with Satan. Djinn are elf-like spirits, some friendly some not, of Arabian origin. They are recognized as beings of a substance other than human. However, like fairies, leprechauns, and the other sprites to which they are related, fun and mischief are their domain. They do not possess the spiritual clout that angels do.

CYCLES OF POPULARITY

Angels reached a height of popularity during the Middle Ages. They experienced an explosion in population and power, being everywhere and protecting all things, and their estimated numbers soared to over three hundred million.

But what goes up must come down – even angels – and various historical events contributed to their loss in popularity. The Inquisition focused attention away from the angels and upon their dark brethren of the Pit. And the Black Plague diminished people's faith in celestial protectors.

By the end of the Renaissance, interest in angels was at a low point as spiritual contemplation was replaced by a new focus on science. The Age of Enlightenment arrived. Knowledge, formerly a privilege of the Church, became more available through the invention of the printing press. Astronomical discoveries put Earth in its place – a minuscule dot in an overwhelmingly large cosmos. Angels began to disappear from serious discussion, and in the art world they were replaced, strangely enough, by pagan gods.

The Protestant Reformation of the sixteenth century rejected the angel hierarchy, with the exception of the fallen angels and their leader. John Calvin denounced the consideration of angels as a waste of time and Martin Luther, while admitting to their existence, suggested that they be kept in perspective and certainly not be considered divine.

The Church has repeatedly vacillated on the subject of angels. Saint Paul denounced the worship of angels, and although in 325 A.D. angels became 'official' in the Church, in 343 A.D. the worship of them was frowned upon. By the end of the eighth century, archangels were accepted to a limited extent, and their names and duties agreed upon.

ANGELS TODAY

Angels have survived the inevitable peaks and valleys in their popularity. While the age of science and information may have pushed angels aside temporarily, they are reappearing as we now shift our focus away from the belief that technology is the answer. The 'New Age' is gradually forcing us to look to alternative dimensions for the answers – be it through hypnosis, yoga, meditation, or channeling – and we need experienced guides to help.

WHY WE BELIEVE

What fuels our belief in angels? Are they a favourite fairy tale that refuses to die? Are they visions witnessed during out-of-body experiences? Are they energies representing our own psychic or higher selves, onto which we project form, making them easier for us to comprehend? Or is it their function as guardians that makes them so universally and timelessly appealing – a security blanket that we can't quite give up?

CHAPTER THREE

THE CHANGING IMAGE OF ANGELS

✦ The ancient Middle Eastern predecessors of the angels were imposing beasts, pillars of ferocious strength. History somehow managed to transform angels into floating beings composed of ethereal light.

✦

Angels gradually lost their beastly forms and became humanoid. Female bodies borrowed from the goddess religions dominated at first, but eventually the angels became distinctly male in form. They lacked wings or halos, and appeared to look just like mortal men.

Early Bible tales of angels depicted them as wingless and rather earthy and humanoid. The three strangers who enjoyed Abraham's hospitality seemed quite unremarkable, until they returned the favour by making his elderly wife fertile. The man-like angel found sitting inside Christ's tomb was also wingless.

However, in 560 B.C. the prophet Ezekiel had a spectacular close encounter with beings that were certainly not human. These winged creatures descended in a fiery cloud, and had four faces each (one human, one ox, one eagle, and one lion). They moved on wheels and made a tremendous thundering noise.

Perhaps such encounters are responsible for the radiant qualities that developed in angel imagery. Halos and shining lights began to accompany them by the end of the fourth century. By the eighth century, the pagan gods and goddesses were again influencing their image, particularly the winged characters, such as Nike, Eros, and Hermes.

But their 'fleshiness' returned with the rise of the Renaissance. During this period – focused as it was on science and the laws of nature – angels lost much of their ethereal, transparent quality, and became more solid. Some even lost their wings. Cherubs were reduced to chubby babies, an image still popular today. Angels were increasingly depicted as feminine, and those that were male were *clearly* male. If they had to lose their wings, at least they grew sex organs as compensation.

By the end of the Renaissance, artists, like the rest of society, began to lose interest in angels. Since then, they never managed to regain their strong foothold in the art world; thus the angelic image established during the Renaissance has remained relatively untouched.

What Are Angels Made of?

The substance (or lack of) of angels has been much debated through history. Some suggest that angels are naturally occuring energies and that if they seem to us to be visible and have form, then it is because we are 'seeing' them with the inner eye, and projecting onto them a visible form that is entirely subjective.

Saint Thomas Aquinas declared that angels are intellect without substance. They are pure thoughtforms. However, they can take on a physical body if they wish and if it makes their jobs easier.

Swedish mystic Emmanuel Swedenborg, in the eighteenth century, discovered that angels can only be seen if they take on a body temporarily, or by being perceived through the inner eye, or third eye.

MALE? FEMALE? BOTH?

Angels are most likely androgynous, representing the ideal union of male and female into one whole. The purpose of the Torah, according to the Hebrews, is to unite the female half of God with the male half of God. Some propose that angels are able to participate (among themselves) in an act of love that, while similar to sex, is really more of a spiritual intercourse – a total union of two souls. The Church states that angels are immaterial and therefore do not have any reproductive abilities. It would also make sense that if they are immortal, reproduction is unnecessary.

Yet angels seem to be able to at least appear male or female. And the Watchers who mated with human women apparently possessed the physical wherewithal to do so. In Genesis 18, the three angels who come to Abraham sit down and eat. However, whether or not the angels actually ate, or only appeared to eat, has been the subject of entire doctoral theses and ongoing debates.

OUR OWN PERCEPTIONS

All of these changing perceptions of angels show how we ourselves mold their image. Angels are vibrations of love and light – they are truly formless. In order for us to be able to perceive them, and to comprehend them, we ourselves project shape, features and identities onto them. We are best able to relate to angels as human-like beings, at least in appearance. They reveal themselves to us in a manner fitting the circumstances and needs of the moment. For example, Joan of Arc perceived the archangel Michael as clothed in golden armour. How appropriate, considering that she was engaged in warfare.

In times of great crisis, we may be so stressed that the appearance of a supernatural being would exacerbate our dilemma rather than alleviate it. Thus, angels appear to us as 'real' human beings – a mysterious stranger who appears suddenly and then vanishes.

A woman I shall call K. C. had an encounter with an angelic mysterious stranger, who communicated an important reminder to her about her lifestyle.

'I was driving back home, a sixteen-hour drive with visits to friends along the way. It was a brilliant day and I stopped at a restaurant for a sandwich. As I was paying my bill, I saw behind the counter, on the wall, a larger-than-lifesize relief of a figure of a sort of happy-go-lucky tramp. It was part of this particular restaurant's theme, meaning the vagabond life on the road. This stylized bum had large, black, bulbous-tipped shoes with a big hole in the toe. As I was paying, I was thinking about how romanticized the "life of a tramp" was depicted, and the chances were than no "real" bum would be eating in a place like that, as it would be too expensive.

'Later, I stopped to get something to eat at a fast food burger place, a quick stop off the motorway. When I ordered, I was the only person there, except for a *tramp* who came in and stood next to me and placed an order. One thing that caught my attention was that he was very soft-spoken and was grinning the whole time.

'He asked if they had pineapple juice. No? Well, they should. And did they have yogurt? No? They should. The young woman behind the counter was amused and slightly bewildered.

'He finally ordered something and pulled out a crisp twenty-pound note! It was very curious. But the one thing that stopped me cold in my tracks was, when I glanced down at his shoes, I saw the very shoe, with the hole in the toe, that I had seen in the earlier restaurant.

'The significance of what he tried to order first was this: at the time, I was going through some medical changes, and I know from dietary readings and research and my own experience that the acid–alkaline balance is very important to our body's system. Pineapple juice and yogurt would help set that balance up. And this "tramp" standing there ordering reminded me of my own health system concerning the acid–alkaline balance, and how yogurt and juice, so to speak, saved my life once.'

Angels also can masquerade as legendary figures and spirits of the dead. Perhaps we would resist communication with a supernatural being – especially if we are not certain such a being really exists – but we would be open to communication with a departed loved one. That, then, is how angels would appear to us. In other words, the angelic force will find whatever medium is necessary to connect with us.

Clearly, our concepts of angels are influenced by thousands of years of historical description and comment, which have imbued certain images into our collective psyche. Yet, as the modern experiences show, angels mutate to fit modern ideas, environments and situations.

Ultimately, if we become receptive to the presence of the angelic force, and its ability and availability to interact in our lives, then less and less stage clothing will be necessary. Perhaps the ideal image of an angel is none at all, but a presence with which we commune via an inner knowing or awareness.

CHAPTER FOUR

GUARDIAN ANGELS

A popular and certainly appealing type of angel is the guardian angel. The idea of a wise spirit looking out for our welfare is a comforting one. Many cultures believe in a type of spirits the Romans called 'manes', or guardian spirits that have developed from the souls of the dead.

The orishas of the Candomblé religion are elementals, or nature spirits, who are assigned in male–female pairs to act as guardian angels. In Spiritualism, and to students of today's New Age, guardian angels are also known as spirit guides.

Philosopher Rudolf Steiner said every person has one guardian angel, who has a complete overview of that person's present and previous lives. This view of the big picture gives the angel knowledge of what experiences are necessary in the person's destiny, and it guides the person accordingly. It is limited in its ability to help, however, by its charge's belief in the angel. If the person even recognizes the existence of the angel, then the wheels of celestial guidance are greased.

The Gnostics' guardian angel took the form of the 'twin angel.' All people, according to this belief, have an enlightened spirit attached to them – their celestial twin. Some Gnostics believed that only very advanced people could communicate with their angel during their life.

Most had to wait until death to be united.

In Christianity passages in the Bible support the idea of guardian angels. In Exodus, Moses is sent an angel to guard him. Psalms 91:11–13 tells us: 'He will give his angels charge of you to guard you in all your ways. On their hands they will bear you up, lest you dash your foot against a stone.'

In the various Christian denominations, belief in guardian angels is strongest in Catholicism. Several movements are devoted to the cult of the guardian angel. Saint Bernard said that it is our duty to love, respect and trust our guardian angels, and this has become part of the Catholic church's liturgy.

Just how influential is a guardian angel? The answer is, as influential as we allow them to be. The more we believe in them and call on them, the greater their presence – and influence – in our lives. Saint Thomas Aquinas said angels are able to intervene in our lives because they possess a vastly superior knowledge of the natural universe, as well as a mysterious power over the material world. Thus, what they are capable of doing seems miraculous to us. In addition, they can act on our senses, inspire us and strengthen our minds, causing us to see our situations more clearly.

Not surprisingly, it has been the leaders of the Catholic Church who have kept the flame of faith for the guardian angel in the face of mounting disbelief and skepticism.

Pope Pius XI (1857–1939) was on particularly good terms with his guardian angel, praying to him every morning and evening – and in between, if a day was rough. Pius XI did not hesitate to acknowledge this publicly, and even recommended the same to others. He confided to the man who someday would be pope himself, Monsignor Angelo Roncalli (John XXIII, 1881–1963), that angels helped him in his many delicate diplomatic dealings. Prior to a meeting with someone whom the Pius XI needed to persuade, he would pray to his guardian angel, recommending his argument, and asking him to take it up with the guardian angel of the other person. Sometimes Pius XI would himself invoke the guardian angel of the other person, asking to be enlightened as to the other's viewpoint.

Pius XI's experiences influenced John XXIII, who also maintained a deep and abiding faith in guardian angels. He used his radio addresses to exhort followers never to neglect devotion to their guardian angels, who stood ready at all times to help. He particularly urged parents to educate their children that they were not alone, but always in the company of their guardian angels.

Pius XII (1876–1958) urged people to renew their devotion to angels. In 1958, a few days before his death, he gave an address to a group of American tourists, in which he reminded them of the existence of an invisible world populated with angels:

Everyone, no matter how humble he may be, has angels to watch over him. They are heavenly, pure and splendid, and yet they have been given to us to keep us company on our way: they have been given the task of keeping careful watch over you so that you do not become separated from Christ, their Lord.

In 1968, Pope Paul VI (1897–1978) sanctioned the Opus Sanctorum Angelorum ('the Work of the Holy Angels'), a movement intended to renew and bolster belief in guardian angels, and to foster a collaboration between angels and humans for the glory of God, the salvation of humanity, and the regeneration of all creation. There are three phases in the Opus.

Many people wonder why, if guardian angels are supposed to protect us as Scripture says, bad things happen in life. Why is one person saved from a tragic accident by the intervention of an angel, and another person is not?

I don't believe that tragedies and difficulties happen because angels desert us. Angels are always with us. For reasons of spiritual and karmic growth, which we may not always be able to fully comprehend, we must at times in our lives go through pain. Some of us do seem to get more than our fair measure of pain, while others of us seem to glide through life blessed by few difficulties. Perhaps these circumstances are decided upon at a very high level of consciousness prior to coming into the world. However, whenever we are confronted with problems, angels are ready to help us cope. Pain provides us with opportunities for tremendous spiritual growth. Angels are not empowered to make decisions for us about our life's plan. They can only intervene whenever necessary to help keep that plan on track. They also can provide a source of spiritual nourishment when we most need it.

A Healer's Angels

Rosemary Gardner Loveday is one of England's most talented clairvoyants, mediums and psychic healers. Rosemary has always believed in guardian angels. She has always believed she is accompanied by her own guardian angel, whose presence she often feels beside her, especially at night when she says her prayers.

Rosemary explained that she distinguishes between guardian angels, other angels, and spirits of the dead, who can take on angelic roles. 'Everyone has a guardian angel assigned to them, for protection and guidance of our souls, and to help make decisions in life. These angels are assigned to come over here and do that kind of work. We're not always aware of our guardian angels. Some people with more developed intuition are more aware than others.'

Rosemary said that one must have faith in the guidance provided by God through guardian angels. 'Faith, in a way, is not knowing what's ahead, and handing over your life to God. You may be asked to do some difficult things, perhaps something you think you can't cope with. But if you have faith, you know that God will look after you, and you will always be protected, whatever the circumstances. Your guardian angel is always there, and you'll always be protected. So don't be frightened, whatever situation you are sent into.'

NAME THAT ANGEL

According to Scripture, there are myriads of angels, yet the only ones named are: Michael, Gabriel and Raphael (the latter appears in the Catholic bible, which includes the Book of Tobit). Other sources however name many angels. When we come into awareness of our own guardian angels, do we have to call them by a name? The early Hebrews believed in a complex hierarchy of angels, all with their own names. In mystical Judaism, names have great significance, and manifest various powers.

Yes, we must discover names for our guardian angels if we wish them to manifest in their fullest magnitude. Naming is an important ritual: it defines, and it invests life, power and potential.

Perhaps the best way to discover the name of a guardian angel is to ask for it in prayer or meditation. The name will arise spontaneously in your thoughts.

✳

CHAPTER FIVE

HOW ANGELS
COME AMONG US

✳ A mystical vision is perhaps the
most dramatic way an angel can manifest
. . . they appear to us bathed in a glory of
light and radiate supreme joy and love.

✳

DIRECT INTERVENTION

Benvenuto Cellini (1500–1571), an Italian goldsmith, sculptor and author, tells in his autobiography, *The Life of Benvenuto Cellini* (published posthumously in 1728) how an angel saved his life in prison. Cellini, a hot-headed man, was constantly engaging in scrapes and fights with other people. On several occasions, he was imprisoned, and was condemned to death. He was absolved once for murder, by Pope Paul III.

In 1535, he was jailed in Rome on charges of stealing the jewels of Pope Clement. Cellini was incarcerated high in the towers of the Castel Sant'Angelo. He made a daring attempt to escape by scaling down the castle walls on a rope made of bedsheets tied together. He was captured and thrown in the dungeon. While in the dungeon, he sank into despair and resolved to kill himself by hanging. Just as he was about to hang himself, a tremendous invisible force knocked him back. An angelic youth appeared to him in a vision and lectured him about the importance of living. Cellini was released from the dungeon on the personal request of a cardinal. He went on to become one of the most celebrated artists of the Renaissance.

Angels also intervene directly in the form of human beings. Typically, these are 'mysterious strangers' who appear suddenly in times of distress. They know just what to do to solve someone's plight. Once the rescue has been made, they suddenly vanish. No one ever knows who they are.

Mysterious strangers can be male or female. Most often, they are male – usually a fresh-looking, clean-cut youth. They are invariably well-dressed, polite and knowledgeable about the crisis at hand. They speak, though they talk sparingly, and they will even take hold of the people in distress.

Mysterious strangers also appear in 'roadside rescues'. Here the stranger arrives to help the motorist stranded on a lonely road at night, or who is injured in an accident in an isolated spot. Or, human beings arrive just in the nick of time.

Jane M. Howard is on the road around the world a great deal to lecture and give workshops. One night, the accelerator pedal in Janie's car became stuck, and she ran off the motorway. She stopped the car by pulling on the handbrake and turning off the engine. It would not restart, and she began to panic. It was 10 pm and she was miles from the nearest exit. She prayed to the angels for help, and within minutes, a van pulled up, carrying a man and a woman. The woman rolled down her window and told Janie not to be frightened, for they were Christians. Even so, many people would have been wary of strangers at night. But the angels gave Janie assurances, and she accepted a ride to a petrol station. She discovered that the couple lived in a town near hers, and knew her family. They pulled off to help Janie, they said, because they had a daughter, and they hoped that if their daughter ever was in distress, she, too, would be aided.

PRAYER AND MEDITATION

Prayer is a powerful way to commune with the Divine. Prayer directs psychic energy toward accomplishment. The noted healer, Ambrose Worrall (1899–1972), who practiced faith healing with his famous wife, healer Olga Worrall (1906–1985), once said that all thoughts are prayers. Thus, we should constantly strive to tune our thoughts to the highest expressions possible of love, benevolence and goodwill.

Angels listen to our thoughts and prayers. If you pray for their guidance, and are sincere and receptive, the guidance will be given.

Carol Ann Durepos told me about a prayer experience she had that evoked a church-full of angels. It occurred during a time when she was coping with family problems and the death of a friend.

'I was able to just settle down and pray when all of a sudden the whole church appeared to be filled up with moving light, flashes and singing. I felt like the angels, saints, seraphim and cherubim were all rejoicing with me and that the very universe, even the stones, sings praise when the name of Jesus is said or spoken. I can't tell you how long it lasted, or whether I saw with my open eyes. It felt like I did see with my eyes – the church was completely full of colour, sounds, and movement, like it was going to burst open.'

Alma Daniel has counselled people internationally on how to reach angels through meditation. 'Sit in a quiet and peaceful place,' she suggests. 'Ground yourself and put your hand on your heart, which connects you to your own loving nature. Close your eyes and imagine the presence of your angel around you. How does that feel? Maybe you feel yourself being enclosed by these wonderful wings. Or maybe you feel yourself growing lighter. Whatever it is, acknowledge what you experience. Just stay with it. Notice how you're feeling, what your breathing is like. It's through the imagination that angels can come to us. If we can conceive of it, it can happen. Then ask the angel for its name. Or ask, "Please be with me," or "Help me in this day." Angels are here to help.'

DREAMS

We have only to look at the Bible for some of the earliest stories about how angels appear to us in dreams. Genesis relates to us the story of Jacob, son of Isaac and Rebecca, who was hunted by his twin brother, Esau. Jacob escaped to his uncle. En route, he had a dream of angels ascending and descending a ladder to heaven. God promised him and his descendants the land upon which he slept. 'I will not leave you until I have done all that I promised you,' God said.

Joseph, husband to Mary, the mother of Jesus, similarly was given important messages by angels in dreams, we are told in the Book of Matthew. While betrothed to Mary, he learned of her pregnancy, and resolved to divorce her quietly. Before he did so, an angel appeared to him in a dream and said, 'Joseph, son of David, do not fear to take Mary your wife, for that which is conceived in her is of the Holy Spirit; she will bear a son, and you shall call his name Jesus, for he will save his people from their sins.'

That angels appear in dreams is significant, for dreams have a reality of their own – they are just as real as our waking consciousness. In dreams, we leave the boundaries of the physical world behind and travel to higher, more subtle planes. These planes are more subjective, but the subjectivity does not render dreams any less 'real.'

Patricia is a born-again Christian who feels in close communion with God, Jesus and angels.

'One night several years ago, I was in my bed at home in a house with my brother and his wife. I felt led by the Word – an inner feeling that I believe is Jesus Christ – telling me to pray that there would be angels around the swimming pool outside. We have a small above-ground pool in the garden. There are a lot of children in the neighbourhood, including my niece and nephew, who go into the pool. So, I prayed as directed. I envisioned angels standing around the pool's ledge facing inward. They looked like Roman guards, ready for protection. I don't know why they had that appearance, that's just how they popped up. I didn't anticipate that there might be an accident. In fact, I kind of questioned why I had been directed to pray that.

'A day or two later, I found myself in the pool with the little children. I don't know why I got in – I usually don't use it. I was standing in the water at one edge, and a friend came over and started talking to me, diverting my attention.

'My sister was nearby but out of sight. She suddenly felt the Lord say to her, "Check on Lisa," her daughter who is two and a half years old. She looked around and realized Lisa was not present. "Patty, is Lisa in the pool?" she called out. I didn't hear her, and she called out again, "Patty, is Lisa in the pool?" I heard her, and turned around just in time to see my little niece step off the ladder into the pool at the other end without her little floaters on. She sank immediately under water. "Oh, My God!" I shouted, and I started running towards her in the water. But the water was heavy, and it seemed like I couldn't get to her in time. Then I heard an authoritative masculine voice, about a foot away from my right ear, say, "Swim to her!" I felt a presence, as though someone were standing near to me.

'I swam to Lisa, and I scooped her out of the water by her arms. Her eyes were open so wide they almost seemed to pop out of her head. By then my sister had reached the pool, and she took Lisa and began banging her on the back. Water spurted out of her mouth, and she coughed and began breathing again. She was all right!'

VISIONS, APPARITIONS AND VOICES

A mystical vision is perhaps the most dramatic way an angel can manifest. Unlike the mysterious stranger, who has the appearance of a flesh-and-blood human being, angels in mystical visions are bathed in a glory of light and radiate supreme joy and love. Such visions usually have an instructional purpose, and a tremendous transformative effect on the person.

Many spiritual leaders have received their initiation by way of an angelic mystical vision. For example, Isaiah was shown the throne of God, encircled by seraphim.

Muhammad (c. 570 or 571–632), the founder of Islam, received the revelations that became the Koran in a series of angelic visions. In 610, Muhammad was forty years old and living a life of asceticism when, one night in his dreams, the archangel Gabriel appeared and gave him the first revelation of the Koran, the holy book that contains the doctrine of Islam. Muslims call this event the 'Night of Power.' For the rest of his life, Muhammad fell into nearly daily trances, during which he received the remainder of the Koran.

Joseph Smith, Jr. (1805–1844), the founder of Mormonism (the Church of Jesus Christ of Latter-day Saints), had various visions, including of angels and pillars of light, that contributed to persecution against him. On the night of September 21, 1823, he was awakened by a brilliant light flooding his room. The light revealed the angel Moroni who revealed to him the location of the gold plates containing the history of an ancient people from Israel. These were translated into the *Book of Mormon*, published in 1830.

Emanuel Swedenborg (1688–1772) travelled often to

heaven in his mystical flights. The Swedish scientist and scholar led a fairly routine life until 1743, when he was fifty-six. The spiritual world was thrust upon him one night in a dream, during which he visited the higher planes. For the rest of his life, he experienced mystical dreams, visions and ecstatic trances, in which he communicated with angels (the souls of the dead), Jesus and God, and visited heaven and hell. The experiences stimulated a tremendous outpouring of written works, which led to the founding of a religion in his name, the Church of the New Jerusalem.

These are but a few examples of mystical visions involving angels. Few people in modern times have such dramatic experiences. Perhaps one reason is that our society does not foster an environment conducive to visions and prophecies. Consequently, our state of consciousness tends to be closed to these types of openings to other realities.

Some modern angel encounters do involve visual

apparitions. The most common forms are columns or pillars of light, or balls of light. No figures or forms are seen in these lights, which usually are a bright (sometimes dazzlingly bright) light of white or silver. The percipient receives mental impressions that these lightforms are angels. Sometimes figures or outlines of figures are perceived within the light. They are invariably human in shape. People occasionally see wings, which typically are as large as the figure itself. It is not uncommon for the lower portions of the figure to fade off into nothingness, or for hands, feet and faces to be poorly defined or not defined at all.

DEATH AND NEAR-DEATH APPARITIONS

Angels stand by us throughout life. When our time comes to leave the earth plane, they assist us across the threshold to the next world.

The appearance of angels, spirits of the dead or other spirit helpers at the time of death has been recorded throughout history. People who are conscious right before death sometimes exclaim on the sudden presence of a radiant deceased loved one, an angel, or another religious figure, such as Jesus or the Virgin Mary.

Occasionally, others can see the take-away angels as well. Shirley Steinhoff Cole told me a story passed down to her by her mother, Louise Regina Beichner Steinhoff, about Shirley's grandmother, Mary Louise Cohlhepp.

'After all her children were grown (nine girls and four boys), all in good health, my grandmother became gravely ill and was taken to the hospital,' Shirley said. 'The doctors told my grandfather that his wife was gravely ill and was going to die, so they put her in what was called the death room. Her sons would hear of no such thing, so the four of them went against the wishes of her doctor, went into the hospital, and carried her out, bed and all.

'When they got home, the girls nursed her back to health. Her hair had even turned white from the illness. When she was well again, her hair came in black as it was when she was young and healthy.

'She was very ill again at age sixty-six, and all the children were notified to come home to see their mother for the last time. They were all there in the room with her, when a band of angels surrounded her bed. Her soul left her body and the angels disappeared.'

Joy Snell, a former nurse, saw take-away apparitions frequently in her work in hospitals. Sometimes she would see angels arrive at the bedside of someone who was on the brink of death, and stand by ready for the moment of transition. Immediately upon death, she would see the soul leave the body and coalesce into a spirit body just above the corpse. It would have the same radiance as that of the angels. Once the spirit body had formed, the angels would take it away.

Snell also saw other apparitions in the course of her

work. One was a ministering angel who took the form of a middle-aged female nurse. At first, Snell thought this angel was in fact a human. She would see this figure moving about wards only at night, simply touching sleeping or unconscious patients who had been in great pain and distress. Invariably, these patients reported the following morning that their pain was lessened or gone, or that they had experienced a profoundly restful night of sleep. Some said they'd had dreams in which they heard the most heavenly music. When Snell inquired who was this fabulous nurse who brought such relief to the patients, she discovered that no one on night duty matched the description.

Two other apparitions Snell saw were harbingers of death and life. The Angel of Death was a dark, shadowy, veiled figure that appeared at the foot of a patient's bed. No matter what the doctors said about the patient's chance for recovery, if the Angel of Death appeared, the patient always died within two to three days. Conversely, the Angel of Renewed Life forecast recovery. This angel appeared at the head of a patient's bed. He was a bright figure in a cloud-like, luminous robe, with a youthful, happy face. He always stood with his right arm raised and index figure pointing up, like a high sign. Again, no matter what the doctors said about the patient's chances of recovery – no matter how bleak – if the Angel of Renewed Life manifested, the patient always recovered.

Angels make themselves known to us in whatever manner and form are the best in order to reach any one individual. If we are receptive and open-minded, some of our experiences may surprise us.

✷

GIVE ME A SIGN

✷ As we've seen in other chapters, angels can intervene dramatically in our lives to save us from certain tragedy. Angels also can give us important information that not only can change the course of our lives, but the lives of other individuals as well.

✷

Most often, angels work in more gentle and subtle ways, guiding us in our thoughts and deeds on a daily basis. These 'little' interventions can have just as profound an effect on our beliefs and life philosophy as angel rescues. Such contact with angels can open us spiritually, steer us onto another course, or simply reaffirm that we're on the right track.

'WHAT SHOULD I DO?'

Leslie of McLean faced a crossroads in life at age twenty-three.

'I was at a very low point in my life. A relationship that I had thought would be permanent was ending, I hated my job, I didn't like my living arrangements, and a general dissatisfaction with my life was causing great despair. One cold November Sunday night I went to church. The auditorium was very large, seating between a thousand and two thousand people. On this night, attendance was sparse, and I sat on next to the last row with no one nearby. Waiting for the service to begin, I remember thinking to myself, "God, what am I going to do?" A clearly audible voice said, "Go back to university." The voice was so clear I turned around to see who sat behind me. To my surprise, there was no one there.

'I had never considered returning to school as a possible response to my situation, but the advice made such good sense that I quit my job, stored my furniture, and within three months was back at university as a fulltime student. The fact that I had only five hundred pounds to my name never seemed like an impediment, and of course, it wasn't. I was able to get loans and find jobs and assistantships all the way through my master's degree. Sometimes the work came in ways that were truly miraculous. My guardian angel seemed to be working overtime.

'The truly amazing aspect of this experience was that I did not think it unusual or strange in any way. I never told anyone of the experience until years later, after I had become involved in a spiritual search. Only then did I truly recognize how I had been blessed, and I was amazed that I had the good sense to follow the guidance. I am very happy that I listened and acted.'

SEALED WITH THE HOLY SPIRIT

If ever anyone needed an angelic high sign, it was Edith. The time was 1972. Edith, a devout Christian, had been struggling for years to cope with serious family problems.

'Have you ever heard of the statement, "Being born of food that has been much tried by fire"?' Edith began. 'Well, in a spiritual sense, it means a person such as I, who has had to go through an awful lot of tragedies and negative situations in life, and in spite of it all, still desiring to become the person the Lord wants them to be.

'As a child, I came from a broken home, with a family

history filled with self-destructive people through suicides, alcoholism and murder. Even though I did feel like an orphan, having to be in and out of a children's home quite often due to family tragedies, I always knew that even though I felt like there was no one else there for me, there was always God.

'Being unfamiliar with the many guises of alcoholism, I ended up marrying an alcoholic in 1968. My then-husband was a *constant* beer drinker, but it never dawned on me that the problems in our relationship were due to the fact that he was an alcoholic. As you can well imagine, I was living through a hellish situation, with his playing all sorts of psychological and emotional warfare to keep a wall between us. Until I came upon Alchoholics Anonymous, I truly believed that the problems were due to me. That was not the case at all, even though my alcoholic husband tried his hardest to make me look like the bad guy all the time.

'You can believe me when I tell you that in spite of all I was going through, I was very close to God. With my own mother committing suicide over my errant father, I kept praying to God to help make me the person he wanted me to be, in spite of everything. I told God that I did not want to commit suicide like my mother did to escape her situation from my father. I prayed constantly to God to *please help me*, even going into the closet and pouring my heart out to him with tears just streaming down my face.

'Then one June night it happened! My husband and I had gone to bed. I was pregnant with my second daughter, who was due in December. I had closed my eyes but had not yet gone to sleep. Suddenly, I heard a voice doing mumbling or chanting of some sort. I immediately opened my eyes, because I thought it was my eighteen-month-old daughter starting to wake up and cry, and I would have to attend to her.

'I was lying on my back and I did not expect to see

what I saw! Above me, there appeared to be someone who looked female floating above me. She was chanting something in a language I had never heard, and all the while was moving her right hand across my forehead. I looked to the right side of my bed, and standing there was what appeared to be someone male. He was wearing a white robe. I quickly looked him up and down. There were long sleeves on his robe, and at the end of each sleeve were two bands of gold trim.

'My eyes kept moving from the being above me to the one beside my bed. The standing one had his hands folded in front of him, his right hand over his left. He appeared to be dark-haired with dark eyes. He had such a wonderful expression on his face, as though he knew something I did not at the time. No words were spoken. I thought to myself, "Are you from outer space?" In just a few minutes, the whole episode was over. The shock of what happened got me out of bed to go think about it.

'After that, I read some books on angels and consulted

with some Christian clergymen, trying to find out what the experience signified. With the spiritual insight I had after the experience, I knew that the Lord had sent those angels as a visible sign to me to let me know that he was sealing me with his Holy Spirit, and the angel above me was writing the Lord's name upon my forehead. I belonged to Him! And, he wanted me to be a witness to the sealing of his Holy Spirit upon me as an assurance to me, because he must have known that I needed that assurance.'

MINDING MORALS

Bob is a retired officer of the U.S. Air Force. He grew up an Episcopalian, and throughout most of his life never gave much thought to angels, though he did feel uncommonly protected and lucky. In the early 1970s, while on a trip for the Air Force to Addis Ababa, Ethiopia, he had the first of some experiences which changed his mind about angels, and had other, far-reaching effects as well.

'I was staying in the Hilton Hotel, and I went to my room. There was a knock on the door, and suddenly there was one of the most beautiful women I've ever seen in my life standing there, very seductively. She had long blond hair and looked like the actress Barbara Eden. I was startled – she was so beautiful. I'd seen her around the hotel and had had some conversations with her. She invited me in her room. I was married – and still am – but I was torn, and was tempted to go. I declined her offer and she left. I wondered, did I make the right choice? Even though I love my wife, at the same time I was certainly tempted. My room had a Bible in it, and I started reading it, and thinking about that aspect of life.

'When I got back home a couple of days later, I was still thinking about all this as I drove home from the base. I hope this doesn't sound silly, but I was questioning, is there a God, and are there angels. As I was driving along, I said, "God, if you're real, give me a sign!" No sooner had I said the words than a brilliant light came right at the windscreen almost into the car. It was like a flash of lightning or a flashbulb going off right in front of me. I was so startled that I almost couldn't believe that it happened. I thought maybe I imagined the whole thing.

'At home, I was greeted by my wife and kids, and I assured myself that I'd done the right thing back in Addis Ababa. I know it may sound obvious to you, but I grew up in a *Playboy* atmosphere where taking advantage of such a situation was considered the thing to do.

'When I went to sleep that night, I was visited by a figure that I can only describe as an angel. I don't know whether I dreamed it, or I awoke and saw it in the bedroom. It was a shining white figure, a huge, tall man in robes, a very loving figure, just radiating love. He walked to the side of my bed and put his hand on me. He didn't say anything, but I got the impression that he was telling me that I had done a good job – the right thing.'

It's the Right Thing to Do

Jane M. Howard is often given guidance by angels that she does not fully understand at the time. However, she long ago learned to trust the guidance, and so she follows it when it is given. Time and time again, what the angels say bears out. Sometimes, what she is guided to do has a great impact on another person, which Jane herself could not have foreseen.

'I received a jewellery catalogue in the post, and in that catalogue I saw a pair of simulated diamond earrings in the shape of dear little angels. I heard my guardian angel tell me that I should buy these earrings. I questioned that message because I have a pair of diamond studs that my boyfriend, Walt, gave me for Christmas two years earlier, and I never take them out. I adore them! I could not fathom that I would ever wear these earrings in the catalogue. However, I heard loud and clear that I was to buy them, and so I did.

'The earrings arrived and sat in their little gift box on my dresser for about a month. Then one day I received a phone call from a close friend. This wonderful woman had just returned from the doctor's office, where she had been told that she had cancer of the uterus. She was so frightened by the bleak prognosis. And she was very confused about decisions she had to make, and she knew she had to make them quickly. She asked if I would pray for her, and I told her I would begin immediately.

'When I hung up the phone at my desk at the advertising agency where I work, and had got up from my chair, I felt myself encircled by the most loving angels of service, who were chiming to me in unison that I needed to go home at lunchtime and package up the earrings – they were for this woman to wear. They were to be sent to her immediately as a gift from the angels with this message: when she wore them, the angels would be there whispering in her ears, "We love you. It's going to be all right. Don't be frightened. We are here with you. You are not alone." The angels added that I was to tell this woman that she was to wear these earrings in "good health," because that was what the angels were sending her – blessings of good health. And so I followed those instructions and the earrings were on their way to New York.

'Several days later, I received a call from my friend. She stated that the earrings had arrived on a day when she was having a battery of exams and tests, and they were truly a gift of strength from heaven. My friend commented that this had been one of the lowest days in her life until the angelic gift of love arrived. She felt the earrings were truly a miracle. She has since shared with me the inspiration that the earrings have given her to face the challenges at this time. The angels constantly reassure her by the presence of the "angelic present" that she is not alone, and to open herself to the gifts of heaven, which include healing – and sometimes even a pair of earrings!'

✵

CHAPTER SEVEN

IN SEARCH OF ANGELS

✵   If you wish to communicate with angels your best bet is Eddie Burks. When I met him he was living and working at Inlight House in Grayshott, Hampshire. The setting was fairytale – a tidy brick country house with a fresh appearance, set on lovely, landscaped private grounds next to a National Trust forest.

✵

Eddie had instilled within the house a soothing energy. At night, especially when the moon rode high in
the sky, the woods and lawn were full of dancing fairies
and nature spirits. It was a magical place, and it was
there that Eddie connected me to angels in a way I had
never before experienced.

Eddie's first peek into the world of spirits came at an
early age; however it was when his wife and son tragi-
cally died that his awareness of angels became height-
ened.

'Margaret – we called her Peggy – came back the day
after she died. My son had come back to the house from
Derby, which was about twenty-five miles from where I
was living, to be with me. We were both feeling utterly
miserable. Lunchtime came and I said we must eat
something. So I took out of the refrigerator a chicken
which I had shared with my wife the day before, and
there was some of it left over. My son was standing near
me, and I was morosely cutting up this chicken, and

suddenly I burst out laughing. My son thought that I'd flipped. I said, "It's all right, it's all right, Michael. Your mother is here and she is reminding me about what happened yesterday." And she was. She was reminding me that I'd cut chicken up the day before, put all the good bits on her plate, scruffy bits on mine, and the two plates somehow got interchanged, and I had all the good stuff and she had all the poor bits. We laughed about it at the time. She was using this event to lift me out of my sadness, you see. From that point on, I felt so different. I was walking on air. She kept coming back and I knew she was around. I knew she had survived. I couldn't see her, but I got a lot of information telepathically from her. That went on quite some time.

'At the time Michael died, he was living in Scotland, where he had gone to get a job. He was living with a girl that he was hoping to marry, his first marriage having broken up. When the lady concerned phoned me to let me know that he died, I went straight up by train. The following morning I went out to get a death certificate. When I went back into the house, she was busy at the kitchen sink. And as I walked into the room, it was as though I had walked into Michael. I said to her, "Good heavens, Michael's here." She didn't even turn round. She said, "I know he is. I've been talking to him." She's psychic, too, but she had repressed it for a number of years. Her psychism was resurrected by his death. Michael was with us a number of times in the days that followed. When I returned five days later to my home in Nottinghamshire, I again felt his presence and continued to feel his presence on and off for quite a time. But I was never thinking that I was bringing him back. He was coming back out of sheer love, I think, because we were very close.'

'At the end of 1984 to early 1985, I was suddenly pitched into healing work. It was spontaneous. As I developed the healing, doors opened in my awareness, and I would find myself involved in other aspects of the person's life.

'For about six years before this, I'd been aware of a presence with me, and it would come to me at any time. I might be driving, I might be watching television, I might be reading, I might be at my office desk. It was benign, and I knew it could be trusted. It brought me a great feeling of uplift and peace.

'I could sense it, but I couldn't hear it or see it. I could sense it, sense it very strongly. It was nearly always on my right side, somewhere behind me. Quite close up, almost like a cloud hovering there.'

The first indication that the spirit had arrived to help Eddie become a healer happened spontaneously.

'One day I went to see a friend whose wife had just come out of hospital after having a severe spinal operation. I went on the spur of the moment. Five minutes before I got there, she got out of bed for the first time since she returned from hospital, put her dressing gown on, and sat in the living room, as though she were expecting me. I tried to start a conversation with her, but

I realized she was too weak. So, I started to talk past her, towards her husband. She suddenly intervened and said, "Have you brought a spiritual presence into the room?" The moment I acknowledged it, I simply had to get up and take her hands, and whoosh! something happened. A surge of wonderful power or energy just flowed straight through me towards her. I'd never had this happen before.

'A fortnight later, at the age of nearly 70, she was dancing on the patio in the back garden to show me how quickly she'd recovered. She was going out shopping. She lived for another seven or eight years. From that day on, I've been coming across people that I have to help.'

One person Eddie helped was a woman who was diagnosed with uterine cancer. He gave her healing prior to her entering the hospital for radiation treatment and a hysterectomy. During the healing, he saw a light shining at the end of a tunnel, and was told that he wasn't to say anything about it.

'I visited the woman several days after the operation. She was sitting up beaming. "The sister is amazed at the way I've come through it," she told me. "But we know why, don't we?" "Yes, we do," I responded. "I've still got to learn the results of the tests they've done to find out whether I have any malignant cells floating about me," she said. But thinking about the light at the end of the tunnel I knew the test would be clear. Sure enough, no malignancy was detected. After several months, she was well enough to return to work. She even learned to drive a car for the first time, and was free to embark on a life full of activity.'

As time went on, Eddie experienced an increasing inner urge to help people with healing. People would come to his house after work, as many as seven in an evening. Eddie did a laying on of hands, following his intuition.

Eddie sometimes received direction from nonphysical beings, without knowing who they were.

'I'm sure there is an angelic presence, and I always invoke an angelic presence. I ask in a silent blessing that I give at the end of each healing for the angelic being most linked with that person to be able to draw close and give guidance and help in whatever way is necessary. I'm much less specific in my healing now, because I've come to realize, as most healers do after a time, that we don't know enough to be specific about how the healing should be applied. Some people need their illness, for example, and it's not up to us to intervene to take that illness away. It's really part of their karmic past, and they must deal with it. So I try to make my healing non-specific, and hand over to whatever presence is with me, including the angelic presence that I always feel is close'.

Eddie had a glimpse of an angel in one healing.

'I started to do healing with the lady who'd learned she'd got cancer, and had to have a hysterectomy. When I went to see her, while I was talking to her, I realized that there would be an angelic presence with her when she went in for this operation. I asked her to visualize this, and to try and draw it closer to her. This I did on other occasions with people who were going in for operations involving general anaesthetic. I felt there was always an angelic presence going with them. Once saw this angelic presence, but not fully. I saw the edge of a garment, and it was a brilliant white. And I knew that I wasn't seeing the whole of this figure, because it would have been too blinding. I came to progressively understand how important it was for people, if they could, to understand that an angel was present. I think the reason for the angel being present, particularly in operations where there is a general anaesthetic, is that the operation is like a near death. The soul is moved out of the body and therefore has to be protected so that nothing else can gain entry. Just as there is an angel present at death, so this angel is present during operations. I used to call this the "Angel of Operations". I couldn't think of any other word for it.

'I made a point when I knew somebody was going to have an operation of always invoking this angel. It would always be there, but I think if the person undergoing the operation could be made to understand that it was there, then the presence could be drawn closer and a lot of help would come out of it. This has been people's experience – in a number of cases they've felt that there was a presence with them. I think angels are present in other circumstances of importance – for example, at birth. I'm sure that they play a major part in that process, and in the process of from conception right up to birth. I believe that there is a hierarchy of angels just as there is a hierarchy of most spirit entities. There is a level in the angelic kingdom where angels overlook houses. I sometimes wonder, if a house becomes empty, that the reason why it falls into decay is as much because the angel withdraws as that it is due to physical decay. A house that's been empty for a long time has a great sadness about it. It's got something lacking. We think the thing it's lacking is, it's got nobody living in it. But I think it goes beyond that. The house has ceased to function in the way that a life has function.

'I've had one stunning experience in relation to angels. This was a case where an angelic being elected to come into the human race. It happens rarely. A lady I knew several years ago was very concerned about her grandson. He was a highly intelligent boy. She is psychic and was aware that he was in touch with the darker psychic side on the astral level. Yet, surprisingly, he seemed to be in control. This was very puzzling. In fact, in terms of normal human experience, he was taking awful risks, and he was virtually playing games with certain entities on that level. He also seemed to be getting himself into all sorts of mischief, including scrapes with the police. He was behaving in such a way that she was concerned he'd turn end up in prison. One night I woke up and I was told that this young man, who was about sixteen, was an angelic being who'd elected to come back into the human race. In doing this, he brought with him some remembrance of his former powers at an unconscious level. It was this remembrance that was leading him into the sort of adventures he was having with the dark side, in which he seemed, surprisingly, to be able to control events. I was also told that at a deeper level he was anxious to get as much experience as fast as possible in the human situation. This was leading him into the sort of behavior that we would regard as very chancy.

'I told the grandmother that I thought it was possible that he would go on getting himself into trouble. That he might even foreshorten his life through an accident of some such, in order to come back after a fairly short interval to learn more in different circumstances. It was as though this being, that was projecting as this boy, had a need to learn as much as possible as quickly as possible. When we looked at his behaviour in the light of this, and what she already knew about him, it did account for the things that were puzzling her, and she accepted this. It helped her in one way that she could see that there was a reason and a logic behind the way he was behaving. The last I heard he was up for several court cases.'

I asked Eddie why it was important for us humans to become closer to the angel realm and if indeed it was at all possible.

'First of all, it is God's purpose that we do cooperate with the angelic kingdom. I think that the angelic kingdom has been waiting a very long time for the opportunity for us to recognize their existence. The angelic kingdom has fallen out of recognition with most people in the Western world, who relegate the whole idea to the Old Testament and the stories there, and think that angel stories don't happen anymore. We think that the angelic kingdom, if it ever existed, served their purpose with God in the Old Testament times. In this way, the angelic kingdom has had to stand aside for most people, sadly waiting for human beings to come round to the realization that many of the problems that have beset us, and particularly those that beset us now in regard to the planet, can only be solved by recognizing their existence and cooperating with them. They have the power, the understanding to help us in ways that we are powerless to work. It's absolutely vital that we recognize this.

'We can reach out to them by transferring the notion of the angelic kingdom at mind or intellectual level down to the heart level, so it becomes a deep conviction. We must work on that conviction. The more the imagination opens constructively to recognize that there is this possibility, then the closer we shall become to angels and the faster we shall be able to transfer what is an intellectual understanding at best into this deeper heartfelt conviction. When we reach that heart level, then cooperation can become real, and we shall begin to have an inner knowing and an inner understanding of the angelic purpose. After all, the power they harness could be used – and should be used – in solving our problems.'

After spending two fascinating days with Eddie learning about his work, I asked him if he could contact the angel realm for me. 'I'd like to ask them for their advice concerning my research,' I said.

Eddie agreed, though he pointed out that he had never before attempted to directly contact the angel realm, primarily because of the intensity of the energy there. He was uncertain that his human consciousness could access such brilliance, but he would give it a try. The session turned out to be one of the most incredible I've ever experienced with a medium.

Eddie closed his eyes and began the process of elevating his consciousness. I closed my eyes and relaxed, soaking up the pleasant atmosphere that saturated the house.

After some time – I was not sure how many minutes had passed – I was suddenly aware of a presence in the room. I opened my eyes but could see nothing out of the ordinary. Yet a presence was there, so powerful and intense that it created a pressure in the room, as though we were plunged under water. I had the feeling that the presence was light that was beyond the ability of the human eye to perceive. In addition, I felt the room become hot, even though the heat was turned down.

The pressure and heat built in intensity. Then Eddie, eyes closed, began to speak. I was able to focus on what he was saying, and write it down. The session lasted but a few minutes, and the pressure continued to build until it became almost painful. Yet, it was a pleasant pain, if that makes any sense. I can perhaps only compare it to the ecstatic pain experienced by some mystics.

The message delivered was this:

'Your interest in the angelic kingdom is purposeful, the full import of which will only be revealed in the course of time. You are being used by the angelic kingdom as a willing and valuable servant to help bring about a greater awareness of the angelic function. Be prepared at some stage to be lifted in your inner vision.

What you see will tax your powers of description. This will be presented to you by those angels working most closely with you. You are not yet ready for this illumination.

'We call upon you to tune into this kingdom not by act of will but by dedication likened to sanctified purpose. Our kingdom is one of great light, and few humans are fit to view it. You will be given the necessary protection.

'As for your work, rest assured that we take sufficient measure according to our purpose. Remember this also: do not reduce us in any way to suit humanity's understanding. This would not be right. Raise humanity instead to meet us through inspired understanding. Impress on people that we are not as they. We are not human. We are no closer to God than you are. But we serve his purpose in a way that you would judge to be more direct. We express a power that you would not understand as yet, except if you experience it. The word you would choose would be "ineffable."

'We differ from you in a number of essential ways. We have no pride. But we have much love. We bear the very essence of unconditional love. Yet, we have the power to direct this and place it where God would have us do so. We seek earnestly to raise the human understanding, to bring it to a sufficient level where the answers to pressing problems can be found. We shall not give the answers – that is not our function. These will come from the devic kingdom working closely with those inspired in the human kingdom.

'So, do you see our function more clearly? In your writing, check what you say against what has been given here, and you will do us a great service, and in turn, humanity. Our blessings come to you now, shafted in intense light, which would not be fitting for you to see, but which will lift you and inspire you in the work ahead. We say farewell with the thought that you and we are likewise children of God doing his purpose.

I was particularly struck by the angels' statement that they are no closer to God than are humans, but just tread a different path. That contradicts what has been handed down to us by philosophers and church authorities. Yet, it made sense to me.

Since then, when I've described the session to others, I've been asked why the angels did not define themselves in terms of 'What is an angel?' In fact, they did, by describing their function and nature. But angels are beyond definition in the human sense, and we waste a lot of time looking at trees and missing the forest. Instead of being preoccupied with what angels 'are' and look like, we should focus on their essence: *unconditional love*.

N.B. Not long after my meeting with Eddie, he departed Inlight House for Lincoln, where he continues to work in healing and releasement.

✷

CHAPTER EIGHT

HEALING HELPERS

✷   If angels make themselves available to help humankind, then surely one of their most important tasks is to heal. There are different kinds of healing – physical, emotional spiritual.

✷

When we are faced with catastrophic or terminal illness, or permanent injuries, sometimes we cannot be cured physically, but we can be healed spiritually. Many persons who work in medicine and health care have had strange experiences that are both wonderful and eerie, in which presences and energies from beyond the physical plane manifest to help the patient. In particular, persons who work in alternative healing, including lay healers such as Eddie Burks who work through laying on of hands or energy transfers, are often acutely aware of presences that guide them in their work. Some call these presences God or Jesus; some call them angels.

The healing ministry of angels is established in Tobit, one of the books of the Apocrypha. Tobit tells how the angel Raphael provides magical formulae for healing. The story underscores the concept that angels do not act on their own, but are emissaries of God, and that they have no physical form, but can create the appearance of form for the benefit of humans.

The story concerns a pious man named Tobit and his son, Tobias. It takes place in the late eighth century B.C. in the Assyrian capital of Nineveh, where the people of Northern Israel have been taken captive. The storyteller is Tobit himself, who is instructed by Raphael to write an account of the events that happen to him, his son, and others.

By his own description, Tobit was a model of piety, walking 'in the ways of truth and righteousness.' He gave money, food and clothing to the poor. He defied Sennacherib the king by burying his fellow Israeli dead, whose bodies were left in the open by their captors.

On one occasion, Tobit, who was fifty years old, was just sitting down to dinner when he learned of another corpse that needed burying. He left his meal and attended to the body. He was defiled from handling the corpse and so did not return home that night, but slept by the wall of the courtyard. He left his face uncovered.

Unbeknown to Tobit, sparrows were perched on the

wall, and their droppings fell into his eyes, rendering him blind. He sought help of various physicians, to no avail. His wife was forced to work to earn money.

After eight years, Tobit, depressed and in despair, begged God to let him die. In preparation for death, he called in his only son and told him to journey to Media, where he had left some money in trust with another man. He instructed Tobias to find a man to accompany him on the journey, and he would pay the man's wages for his time and trouble.

While this drama was unfolding in Nineveh, another was taking place in Media. There, a young woman named Sarah was possessed by the demon Asmodeus, 'the destroyer.' Sarah had been given to seven men in wedlock, but the demon had killed them all on their wedding night, before the marriages could be consummated. Sarah's parents, Raguel and Edna, feared they would never marry off their only daughter.

God heard the prayers of both Tobit and Raguel, and

dispatched Raphael to heal Tobit's blindness and exorcise the demons from Sarah.

When Tobit went looking for a man to accompany him on the trip to Media, he found Raphael, who appeared as a human and introduced himself as Azarius, the son of one of Tobit's relatives. They struck a deal for wages and departed.

The first evening, they camped along the Tigris River. Tobias went down to the river to wash, and a giant fish jumped up and threatened to swallow him. Raphael told him to catch it, which he did with his hands, and threw it up on the bank. Raphael said, 'Cut open the fish and take the heart and liver and gall and put them away safely.' Tobias did this. They then roasted and ate the rest of the fish.

Tobias asked the angel what use were the saved parts.

Raphael replied, 'As for the heart and the liver, if a demon or evil spirit gives trouble to any one, you make a smoke from these before the man or woman, and that person will never be troubled again. And as for the gall, anoint with it a man who has white films in his eyes, and he will be cured.'

As they neared their destination, Raphael told Tobias that they would stay in the house of Raguel, and that he should take Sarah as his wife. Understandably, Tobias was not thrilled to learn that seven prospective husbands had all died at the hands of the demon. But the angel assured him, 'When you enter the bridal chamber, you shall take live ashes of incense and lay upon them some of the heart and liver of the fish so as to make a smoke. Then the demon will smell it and flee away, and will never again return. And when you approach her, rise up, both of you, and cry out to the merciful God, and he will save you and have mercy on you. Do not be afraid, for she was destined for you from eternity. You will save her, and she will go with you, and I suppose you will have children by her.'

The events came to pass as the angel predicted. Tobias was offered the hand of Sarah in marriage, and a contract was drawn up immediately. In the bridal chamber, Tobias followed Raphael's instructions for exorcising Asmodeus. The demon fled to 'the remotest parts of Egypt' (the traditional home of magic and witchcraft), where Raphael bound him up.

After a fourteen-day wedding feast, Tobias, his bride and Raphael returned home to Tobit. Tobias anointed his father's eyes with the gall of the fish, and Tobit's sight was restored. In gratitude, he and Tobias offered Raphael half of the monies that Tobias had retrieved from Media.

The angel then revealed his true self to the men. 'I am Raphael, one of the seven holy angels who present the prayers of the saints and enter in the presence of the glory of the Holy One,' he said. He told Tobit that he had been ever present with him, and had taken his prayers for healing to God. He urged the men to praise and thank God, and to lead righteous lives.

Tobit and Tobias were alarmed to be in the presence of an archangel, and fell to the ground in fear. But Raphael assured them no harm would befall them. 'For I did not come as a favour on my part, but by the will of our God,' he said. 'Therefore praise him forever. All these days I merely appeared to you and did not eat or drink, but you were seeing a vision. And now give thanks to God, for I am ascending to him who sent me. Write in a book everything that has happened.' And with that, Raphael vanished.

Raphael continues to minister to healing needs, taking the earnest prayers of humans to God, and responding when assigned. Perhaps he is assisted by a host of other healing angels as well, for those who work as healers attest to a variety of helping angelic beings. Are they other angels, working under Raphael's supervision? Or are they Raphael himself, manifesting in whatever guises are required?

ROSEMARY'S ANGELS

Rosemary Loveday (whom we have already met briefly) is a talented English clairvoyant, medium and healer who radiates a gentle angelic energy. She has been clairvoyant from childhood – a gift passed down through the family on her mother's side – and has been cognizant of angels and souls of the dead for most of her life. For Rosemary, the boundaries between our physical world and other realms are often thin or transparent. From these other realms come divine guidance that provide her with extraordinary gifts. Her work has gained her increasing attention on both sides of the Atlantic. She has helped many people in their struggles with physical illness and other difficulties in life. People often report that her very presence is soothing and healing. When they compliment her on the beneficial effects of a reading or healing, she demurs, saying, 'It is God working through the angels.'

I met Rosemary in 1991. At the time, she lived in the little fishing village of Brixham, tucked onto a cliff on one side of Tor Bay (she has since moved to Plymouth).

I was greeted by a tall woman with a serene and beatific presence, and a soft, angelic voice. I relaxed and knew that I was in good hands.

Rosemary escorted me into the cottage. It was warm and inviting. Built in 1640, it featured three small rooms downstairs: a sitting room with fireplace, a dining room and a kitchen. The low ceilings had exposed beams. There was an atmosphere of peace and tranquility, which seemed to be the product of Rosemary herself.

Rosemary fixed me a cup of tea, and we sat down at a little wooden table in the dining room. I handed her my watch, and she proceeded to give me a reading.

Over the years, I had been to many psychics, mediums and channelers, and seldom did I receive what I considered to be a good reading. Essentially, I believe that we can get our answers ourselves, by developing our intuition and tuning ourselves to the inner voice, which delivers information from a higher source. Readings by others can be helpful, however, when we are too close to problems to see the forest for the trees.

The reading that Rosemary gave me was remarkable. It was accurate, but accuracy alone only validates the facts. More important, her reading was permeated with a sensitivity that was as though she truly were seeing things from my eyes. It gave me lots of new insights, and I felt much better.

After the reading, Rosemary gave me healing. She said a silent prayer for guidance, and then placed her hands on me where directed to do so. I felt surrounded by invisible but loving presences, and heat and energy flowed into me, like I was a battery being recharged. Rosemary told me later that during the healing, she had a vision of a meter, like a petrol gauge, rising from empty to full. It certainly expressed my own feelings of having been refreshed and restored.

Rosemary and I became good friends, and have since visited each other. On one occasion, I had what I considered an unusual healing of illness from her.

It took place in January, when the weather was cold and damp. I arrived in Brixham feeling a severe respiratory infection seizing hold of me. I was badly congested in head and chest, and feared that my entire stay with Rosemary was going to consist of lying in a sickbed.

Rosemary offered to give me a healing. I agreed, though I doubted that anything would help. Sure, it might give me an infusion of energy, but cure me of a cold or bronchitis? I've had countless respiratory infections in my life, and recognize the signs of being beyond staving it off – after symptoms are advanced to a certain point, I just have to ride it out for several days.

After making her prayer for divine guidance, Rosemary placed her hands on my head and then at the base of my throat. These positions were held until she received guidance to cease. Then she took me into the sitting room, where a fire crackled in the fireplace, and tucked me onto the sofa with instructions to take a nap. When I woke up two hours later, most of my congestion was gone, and I was well by the next day. I was amazed. It was the first time I'd ever been cured of something by a laying on of hands.

A cold is one thing, and major illness is another. Rosemary has aided people with serious disorders. I hasten to add that she does not diagnose, nor does she advocate alternative healing as a replacement for standard medical treatment. The people who come to see her are guided, she believes, by angelic hands, and they seek her kind of healing in addition to other treatments. Perhaps most of all, they seek the angelic comfort that emanates from her.

Rosemary feels the constant presence of angels in her readings and healings. 'When I'm doing my healing, I feel angels draw close,' Rosemary said. 'They help me with my healing. I'm just a channel. They draw close to give people peace. Peace and love come foremost from God, but he sends angels out to gather around people. I handed my life over to God to help heal people. I think if you hand your life over to be used in that way, he's going to send you help.'

Prior to engaging in her work as a clairvoyant, medium and healer, Rosemary worked in hospitals as an aide. She often was aware of the presence of angels attending to the sick and dying.

On one occasion, while working at a hospital in Hawkhurst, she escorted a group of patients on an outing to the theater. During the performance, an elderly man collapsed of a heart attack. Rosemary helped him into a wheelchair and called for an ambulance. His condition deteriorated so rapidly that she had to get him out of the chair and onto the ground so that she could do mouth-to-mouth resuscitation. When the ambulance arrived, Rosemary sat in the back with him and held his hand. He was unconscious, but she had a sense that being with him and holding his hand provided him reassurance. She was also aware of angelic presences accompanying them on the trip. They were very close, and closing in tighter as the ambulance travelled. She closed her eyes and concentrated on providing comfort.

The man died before the ambulance could reach the hospital. Rosemary was still clasping his hand when one of the attendants opened the ambulance door and gently took her hand away, 'He's passed over now,' the attendant said.

Three nights later, a presence beside Rosemary's bed awoke her from sleep. It was the elderly gentleman. Rosemary could see him plainly, as though he were still alive. He was radiating warmth and love. He had come to give her a message, which she received as a thought impression.

'Thank you,' he said.

Spirit presences invariably manifest when she does a reading or a healing. Some are souls of the dead, such as family members of the client. Some are guardian angels attached to the client, and some are angels who have come to assist with the healing.

The angels are accompanied by a pure, beautiful aura of silver-white light. It is a healing light, and it manifests as Rosemary places her hands on the client, and seems to flow through her hands into the person. She suddenly becomes aware of an angel – or angels – standing beside her and the client. The awareness is through the inner eye, for Rosemary works with her eyes shut.

'I do see them,' Rosemary told me. 'They appear not as a man or a lady, but in a sort of long white robe with fair hair. They have a white aura all around them. I find that if I tell the person that they're not alone, but that angels are always with them, it helps them a lot. Angels are something that they can identify with. I think most people want very much to feel that there is a spiritual presence close to them.'

Once a woman came to see Rosemary fraught with anxieties over situations in her life. Rosemary gave her a reading and healing, and felt the strong silvery-white light flow through her. The woman, who was sitting with her eyes closed, said, 'It's strange, but I can feel an overwhelming sense of peace.'

'There's an angel right by you, that's why,' responded Rosemary.

'An angel?'

'Your guardian angel,' Rosemary said. Then she opened her eyes and looked at the woman. To her surprise, the woman's face was illuminated and transformed. When she had arrived for her appointment, her features had been pinched and drawn with anxiety. Now they were full and serene, and a beautiful, angelic expression graced them.

Since then, Rosemary sees facial transfigurations often during healings. 'I think the whole basis of healing is to touch the soul,' she said. 'We're brought into this life, basically, to find God and to find ourselves in a spiritual life. When one gives healing, it's a direct link with God and angels – who are superior beings – to touch the soul of the person so that they start to grow in a spiritual way. If a person lives a materialistic life, at the end of the day, their soul passes on and they can't take their material things with them. So, if they don't develop in a spiritual way, it's a wasted life. Then they have to come back again to another life.

'If I can give a person an inner peace, and also show them that they're not alone, that's one of the greatest gifts I can give. I give healing with that intent first and foremost. Then I'm helping to heal the mind, and sometimes the physical heals from that. Our Lord Jesus is the finest example of a healer. He healed spiritually, healed minds, and healed bodies, as well.'

I asked Rosemary to describe some of her healings with others.

'A lady came to see me who suffered from a loss of

balance, and congestion down one side of her head. I gave her healing, and she became very peaceful and serene and said she felt a lot better. Two days later she rang me up and said, "You know, I'm feeling so much better, you put everything right." Then her husband came to see me, and they both came back to see me. He said he was very pleased what had been done for his wife. I thought it was the lovely angelic presence that achieved that, not me.

'Another lady had a lack of mobility in her hands, and lumps on her knuckles. She was going to have an operation in hospital. I said a little prayer to help her, and gave her some healing. I was guided to work from the elbows to the wrists to the hands. Her hands were red. Afterwards there were little dents on one hand where the lumps had gone down, and both hands were a nice pale colour, instead of being red. The mobility came back, so that she could use the fingers.'

Rosemary went on, 'Another lady came to me for a

reading. She said, "I feel I am at a turning point in my life, meeting you. It has made me look at everything in a completely different way." Readings really help someone find their pathway. People often get to a stage in life and think that there's got to be more to life than this, or which way do they turn – they're searching. The reading helped this lady onto the right pathway. I also gave her some healing for back problems she'd had for years. Later she rang me up and said that two days after having the healing, her back problems were gone, like a miracle.

'Another lady came in a state of depression and anxiety, really awful turmoil. After healing, she said she had a feeling of peace. Although we couldn't change the circumstances of the difficult marriage that she was in, the healing did help her cope. She came back to see me two weeks later and told me she felt as if I'd been with her the whole two weeks. "You didn't just pick me up and put me back on the pathway, you walked around by the side of me and held my hands," she said. She explained that she went home from the healing, and because she felt so different, she was different with her husband, and in turn, he responded to her, and in turn, their daughter was happier. So the whole family unit became happy.

'I think everyone has a little candle inside of them,' Rosemary said. 'I helped to light one in her, and then she went home and lit another one in somebody else, and it spreads. It doesn't just stop at the individual healing. That person will be different, and it will either spread through their family or through their friends.

'You know, it's nice to physically heal. So you make somebody's arm better. Perhaps it just stops there. But people can cope with their physical ailments. For spiritual benefit, they must understand the reasons why they've got them, and the lessons they've got to learn as a result. Then they can basically go on and spiritually help other people. I've had a lot of people come to me who have had healing gifts handed to them, and I've been able to help them, send them out to do more good works. From one little candle, you can light a lot of lights.'

Rosemary said she thought healing could be accomplished in many different ways, and that everyone has at least some ability for healing others. 'I think you can heal people by talking to them, even on the phone. Sometimes by just being with them. You can write them a letter. I don't think you always have to lay hands on somebody.

'People who have come to see me, and are facing difficulties in life, have often said to me they could hear me talking to them in times of need, as though I were actually there with them. I think the angels take the healing to the people, wherever they are, as it's needed, even the night while they sleep. It's one of their many missions.'

CHAPTER NINE

SPIRIT POWERS OF THE NATIVE AMERICAN WAY

✷   Kenneth S. Cohen, also known as Ken Bearhawk, is a renowned healer who lives in a log cabin in the mountains of Colorado. Ken is trained in Native American and African healing and the Chinese technique of QiGong.

✷

In his work with Native American healing methods, Ken is aided by animal totems, or spirit guides or helpers. Totems are roughly equivalent to the Judeo-Christian concept of angels. Their purpose is to help the spiritual development of a person, and to provide sources of spiritual and psychic power. They are considered to dwell literally within the body, rather than exist externally. Unlike the guardian angel, who comes in at birth and stays until death, totems change during the course of a person's life, depending on spiritual development and needs.

In the Native American tradition, there are various ways that a person can find his or her path of power. One of these is hereditary. Another is through transferral of power by an elder or medicine person. Another is through training. Ken is not Native American by birth, though he feels a profound connection to the Native American path. He has developed his healing power for more than twenty years by training and transference.

He was apprenticed to the Cherokee spiritual teacher, Keetoowah Cristy, the great-grandson of Ned Cristy, a Cherokee hero. Keetoowah Cristy transferred power into Ken's body by using a quartz crystal, which became part of his blood, his being. The use of crystal to transfer medicine power is common throughout shamanic cultures.

I asked Ken to describe his healing methods, and to tell me about some of his experiences with totems.

'In healing, the first thing I do is counsel with the person to get a sense of their situation. That way I can determine whether I have an ability to help them. A good healer knows his strengths and limitations. There are people who come to me who I cannot help, and I make referrals. If I determine that they have a problem that I can address, especially if it is spiritual in origin, then I wait a couple of days and see if I receive guidance from my spirit helpers. Then I meet with the person again, and start. I consider everything I do as complementary therapy instead of alternative therapy.

'In working with patients, first I purify them by smudging, which is using the smoke from burning sage. Some use sweet grass or cedar. I might see more about their illness through the smoke.

'I use a lot of prayer, opening myself to a transpersonal source of power. I feel that when I do healing, it is not coming from me, but is moving through me. I am able to connect to the Great Mystery, the Creator, who does the healing.

'I do a lot of noncontact healing, in which I visualize, energy passing through my hands to the person. Occasionally, the healing work will take the form of my placing sacred stones on their bodies. The stones have their own healing natures, and are tuned through my prayer.

'There are various spiritual techniques that might be required in particular cases. For instance, I might see that a source of a person's problem is that they are out of touch with their spirit guides, their totems. If I sense that is the case, or the spirit helper has left, or the connection to the person is very tenuous, I perform a healing ceremony to re-establish contact. That can be done through prayer and calling songs that invoke the presence of a particular helper. It can also be done by waving the person's spirit helper back into the body with a feather fan, or using the breath to blow the helper back into the body. Occasionally I've asked a plant, such as tobacco, which is sacred, to help me with this. I take the tobacco smoke in my mouth and blow the spirit helper back to the body with the smoke.

'I feel that spirit helpers and angels are basically the same. However, in Native American culture, there are different ways to conceptualize the helpers, all of whom are messengers for, and aspects of, the Great Mystery. In the Seneca Wolf Clan Teaching Lodge, Grandmother Twyla speaks about four spirit band members, who are guides who tend to appear in human form. My own visions tell me that these four band members represent the four directions, each with its own teaching. A person might not recognize or work with all of them at the same time. You might go through certain experiences in your life that help you to become aware of one particular band member, for example, the east, which is the direction of inspiration and creativity, and having the wide, panoramic vision of the eagle. After some time, you might become aware of the band members in other directions. It could take a year, it could take ten years, it could take twenty years to learn all four – it's an individual process.

'In my work, I've discovered most people have two totem animals, who are in addition to the human-like four band members. The totems represent the conscious and the unconscious – the obvious aspect and a more hidden aspect of one's self. All – totems and band members – are sources of knowledge and wisdom. One can learn to dialogue with them, in a way analogous to the process of creative imagination. They are also actual sources of power, because they are intermediaries between the physical world and the Creator, the Great Mystery. They can be sources of power channeling. Perhaps "funneling" is a better word. They funnel power down to us when it's needed. In shamanic cultures, if you lose touch with your power, whether it be a totem or a band member or some combination of the two, then you are vulnerable to misfortune and disease. A healing is necessary to re-establishing that contact, and restore power and vitality.

'One can lose touch with their power through not following their original instructions – the assignment that the Great Mystery has written in each of our hearts. In other words, one may lose touch by not having the courage to be fully themselves and to follow the promptings that come from inside, and instead are false to their medicine. For instance, one is not true to their medicine if they live their lives according to other people's expectations, or the conditioning influences of educational and religious institutions. One can also be disrespectful to their medicine. For instance, if someone has a particular gift to be an artist, or poet, or singer, and they don't allow themselves to express that gift, they are disrespectful to their medicine, and they become sick.

'A good example is a woman I worked with one time. I felt she had a great deal of sensitivity to people and their feelings, and that this could have been turned into a real advantage in her life. She could have been a wonderful therapist. But she had so little confidence in herself because of an abusive childhood that she never allowed herself to follow what she really felt should have been her authentic lifestyle or career. So she took up a job in business and became a manager. She didn't hate her work, but she always had an unfulfilled need to work with other people. Her sickness was chronic fatigue syndrome, which she has suffered for many years.

'When I looked at her, I saw that she had a deer and an eagle around her. Very clearly. This made sense to her

– these were symbols that were relevant to her. I told her the deer was her sensitivity to energy, and the eagle was her penetrating vision or ability to look through the shells, through the masks that people use to project their social roles. She could see through that to what was authentic underneath. She began to work with those totem powers and as a result, the chronic fatigue syndrome lessened its hold on her. So, working with the spirit powers can have a very concrete effect on a person's personality and health.'

'Were her totems comparable to guardian angels?' I asked.

'Yes, although totems change in the course of one's life. Sometimes new spirit powers will seek to come in to further growth and development, but a person will not make the life changes necessary to allow them in. The person will sometimes suffer an illness, which is due to the fact that the old spirit power has left, but they have not made themselves fit to receive the new spirit power.

'To sum it up, the spirit powers are helpers and messengers, and they usually come to us as a grace. Our training can invite their presence, but we can't specifically stalk or track a particular power. We can only make ourselves open to it. What my elders have always told me is that the most important power, the highest power, and the one which people must always keep central in their lives – especially if they're working with the spirit powers – is the Creator, or God.'

'How do we open ourselves to the spirit powers?'

'The first thing is to find out what animal you have an affinity with. Some people just know that. They know that because of their dreams, or because of feelings they have when they see a certain animal – they feel like singing and dancing. If a person does not know intuitively what his totem animals are, it probably means he has not spent enough time in nature. There is no substitute for that. It's not enough to simply do guided imagery work and to imagine a situation where an animal can come to you. I feel a person must physically go out and be in nature. When you live close to nature, then these spirit animal presences, as well as the physical animals, make themselves known.

'I once had a client who felt victimized at work. Her boss didn't like her, and she felt her boss tried to turn other people against her. She just couldn't stand it, and she wanted to get a transfer to a different office. But that was up to her boss, who hated her anyway, and she just felt trapped. When I asked her what animal she felt closest to, she said cats. As soon as she said it, I saw a mountain lion. I asked her how she felt about mountain lions, and she said she loved them. I knew I was on the mark.

'I said, "All right, the mountain lion is a hunter – it is not the hunted. The mountain lion can be invisible. When you go to work this week, I want you to imagine that you are a mountain lion and that way you will not be a victim. Your boss will not notice you, or your boss will be so scared of your power that he will not say anything bad against you. Let me know what happens as a result."

'I saw her again two weeks later. She told me she had been transferred to a different office. During the time that she was doing her visualization as a mountain lion, her boss did not notice her and did not make any rude remarks. She felt a radical difference in her health and her spirit. She was very happy with the new place and new employees she was working with.

'It's not only important to invite the spirit presences in, it's also important to honour them and express gratitude to them. Honouring and gratitude are the reasons for Native American songs and dances. You can go out into nature, some place where you can be alone, and sing a song to your animal, dance it, or move like it. Some people draw a painting as a way of honouring their totems. Find some way to express gratitude, because any power or gift that you are grateful for then becomes stronger in you.'

I thanked Ken for sharing his experiences. I saw many parallels in the ways that angels work and the ways that spirit helpers work. They all come from the same source, God, or the Great Mystery.

'Working with angels or spirit powers should not be the object of one's spirituality. These are merely helpers, and one's devotion has to be fixed on the Great Mystery, the Creator. If I meet someone who speaks too much about powers or presences, I feel they've been side-tracked, or perhaps their motivation is not right. Perhaps they're more interested in power or in the emotional thrill they get out of working with these different helpers. One's dedication really has to be centered all the time on the Great Mystery.'

✴

CHAPTER TEN

A PATH OF LIGHT
IN DARK TIMES

✴ Our belief in angels is on the increase. Spiritual emptiness, and a need for more personal spiritual relationships has created a collective desire for angels, and encounters with them.

✴

Angels are also a part of our mythology, and we are attempting to reclaim the riches of our mythological tradition, which have been buried by science and technology. We are in need of great help because we have great problems, on both individual and global scales. And, interest in angels follows a more general trend over the past fifty years or so of increased openness to the paranormal, and an increased willingness to discuss paranormal experiences.

One important reason we are having more angel experiences is the 'consciousness revolution,' that is, a collective uplifting and expansion of human consciousness to higher levels, which provides us easier access to nonordinary realities. We are breaking through to these nonordinary realities with increasing regularity. This is due to three factors:

1 *The overall increase in world population, hence more anecdotes*

2 *An increased willingness to disclose exceptional experiences*

3 *An evolutionary advance in our collective consciousness that is pulling us into other realities, which is the expression of the true nature of our unlimited, multidimensional consciousness*

ANGELS OF DARKNESS

So far, I have not talked much about the dark angels – those who fell with Lucifer to become the demons who try to lead us astray, away from the light of God. Dark angels do exist. Just like the angels of light, they are among us. And, just like the angels of light, they can assume many forms and approach us in many ways. They may be the ugly demons described in medieval writings, but more often than not, they approach us in the guise of humans so that we let down our guard. Like angels of light, they can appear as mysterious strangers.

They can come at us through other people – individuals whom we trust, but whose various weaknesses allow the penetration of darkness into their beings, and who may act unwittingly in carrying out the intent of the dark forces.

The purpose of the dark angels is at least to neutralize us as centres of light and love, and, if necessary, to destroy us – spiritually, psychically, even physically. They are engaged in a spiritual warfare with the forces of light for control of our souls. The more we can be encouraged to petty acts of meanness and falsehood, to major acts of destruction and violence – the more we empty our souls of light and fill them instead with darkness – the stronger the dark angels become, and the further away we fall from God. The choice is ours, because both the forces of light and the forces of darkness present themselves to us. We choose which one we want to follow. And when we choose darkness, the angels of light cannot rescue us unless we realize the folly of our choice and turn to them for help. When we ask, even with the smallest cry of despair, the smallest prayer, the help is given in the greatest abundance and love.

The angels of light and dark watch each soul. The dark ones have a harder time getting recruits, for there are not many who consciously dedicate themselves to the path of darkness and evil. Rather, some people fall onto that path through temptation, and thus they are recruited indirectly. Most people want to lead righteous, good lives, and so they pledge themselves to the path of light.

In my sessions with Eddie Burks, I asked him what he thought about dark angels.

'I think that just as we have dark spirits among us, we have spirits that fall away into dark regions when they die, not permanently, but sometimes for a long time. Then there are corresponding angelic beings who are operating at that dark level. I believe in the existence of the archangel Michael and of the fallen angel Lucifer.

I once had a most extraordinary encounter with the dark angel Lucifer, or a representative of his. I saw a figure and he had a black cloak on. Its face was extraordinarily handsome, and it had curly hair. No horns, I hasten to add. His face was dark, not like a black man's face, but it was dark and it shone – it was reflecting light. And as I watched it, it opened its cloak so that one could see the lining within the cloak, and the lining shone a brilliant silver light. I was told that Lucifer longs to show his other self.

'I believe that what [Lucifer] does is to bring the idea and the reality of evil into the world to give us an alternative path to follow. In offering us this choice between good and evil, he is making it possible for us to gain experience of the dark side, and to gain experience in depth in more than one sense. We gradually learn, over many lifetimes, to climb out of that dark pit that we dig for ourselves. Having climbed out of it, we've learned a great deal, we've become much wiser. We have a wisdom that couldn't come about simply from purity.

'I think Lucifer's function is to be continuously offering us what the Bible refers to as temptation, but I tend to think of it as the alternative power. At every stage in our lives, we're offered this other path. Without this notion of evil, without his presence, we humans would be much more like the angelic kingdom. It is through Lucifer's intervention that we get ourselves in a mess, and that we learn through our own suffering.'

EVALUATING ANGEL ENCOUNTERS

An encounter with an angel does not truly benefit us unless we have a way of interpreting it so that it makes sense to us in terms of our worldview, and we can integrate it – that is, accepting it as a real experience. If we are ambivalent about angels, and one manifests to us to rescue us from disaster, we gain nothing beyond the luck of the rescue if it does not alter our beliefs about angels, and in turn our beliefs about ourselves, our souls, and our relationship to God.

Encounters with angels are part of a broader spectrum of unusual experiences that incorporate altered states of consciousness, nonordinary realities, psychic phenomena and elements of the mystical.

For guidance on how to evaluate our experiences with angels, we cannot necessarily turn to religion or science. Modern religion is rather uncomfortable with the question of angels; science tends to disregard or deny that which cannot be measured, quantified or qualified. We are therefore left to our own devices – our own worldviews, our own cultural beliefs, our own religious beliefs – when it comes to accepting and interpreting the existence and activities of angels, as well as any other exceptional experience that might happen to us.

Angels give us something positive and useful, whether it is information, inspiration, or reaffirmation of self-worth. They bring positive, not negative, experiences that open up doorways to life, or doorways to other realities – or both. They connect us to the highest and deepest parts of ourselves, to each other, to all things, and to God.

Believing in angels, or thinking about angels, especially if one has had angel encounters, encourages more experiences of the same sort. Philosopher Michael Grosso calls this the 'mirror factor of the psychic universe.' The psychic universe reflects back to us what we believe, and we then experience what we believe. The stronger our beliefs, the more feedback we get in terms of experiences.

I believe that angels resort to drama only when necessary, in order to break through our barriers of consciousness. Once they have broken through, they can commune with us on more subtle planes. All of the persons who'd had dramatic experiences did feel continually in the guiding presence of angels thereafter.

We may wish for drama – don't we all love good entertainment? – but drama is not required in order to meet with angels. In fact, the angels would prefer not to use a heavy hand. They would rather not crash about in the material world. They would much rather have us attune ourselves to the frequencies of a higher consciousness, to the true music of our souls. When we do that, and we listen carefully, we can hear the angels singing. They give us their songs of guidance. They sing God's love.

Everyone possesses the gifts for communing with angels, whether it be through visionary experience, clairaudience or the intuitive voice. 'You have to want the communion, and make the commitment to the angels to work with them,' said Jane M. Howard. 'Most of all, you have to be responsible. What are you going to do with the results? If you're sincere about growing and helping other people, the angels will be glad to work with you. Expect miracles.'

One can invest too much in angels, however. We can turn them into idols, which would upset the balance of our relationship with them. Or, we can also trivialize angels with pop culture cuteness. Angels are not dolls or pets. They are awesome beings deserving of great respect. As the angels told me in that session with Eddie Burks:

> *Do not reduce us in any way to suit humanity's understanding. This would not be right. Raise humanity instead to meet us through inspired understanding. Impress on people that we are not as they. We are not human. We are no closer to God than you are. But we serve his purpose in a way that you would judge to be more direct.*

A number of persons interviewed for this book supported the belief that angels cannot help us unless we invite them to do so. In some cases, the individuals were told that themselves that by angels. Or, it was not until they faced a crisis that they asked for help. This need to ask first is a general belief in occult mythology: that one cannot do business with the spirits, whatever they are, without first offering an invitation to cross the dimensional threshold.

We can ask for help in many ways. Sometimes, we ask for help unconsciously, that is, through our Higher Selves. We can facilitate the interaction with angels by direct invitation, especially through prayer, meditation and visualization. And while we can ask for ourselves, we will achieve far more if we ask for help for others – our angels working through the angels that protect other people.

Always give thanks first. Give thanks for the blessings in life, and the opportunities that lie ahead. Give thanks for the guidance that is being sought. It will be given.

Remember the words of the angels:

We bear the very essence of unconditional love.